T0076280

Praise for Samuel T. Wilkinson's

Purpose

"Psychiatrist Sam Wilkinson digs deeply into recent insights about how evolution has shaped the competing dispositions of human nature, and how these observations point to a Creator God who has a purpose for our existence. If you are one of many in our technological society who is troubled about whether science and faith can be harmonized, you will be reassured and inspired by this intellectually rigorous and spiritually compelling presentation."

—Francis S. Collins, MD, PhD, leader of the Human Genome Project, and author of *The Language of God*

"An essential book by every measure. Beautifully written, superbly researched—and life changing. You will never think of your life, or the earth, or the purpose of each in the same way again!"

—Greg McKeown, *New York Times* bestselling author of *Effortless* and *Essentialism*

"If you struggle to reconcile faith and reason, Sam Wilkinson's profound book *Purpose* was written for you. You will be left with an understanding of the guiding forces behind human evolution and behavior."

—Arthur C. Brooks, professor, Harvard Kennedy School and Harvard Business School, and #1 *New York Times* bestselling author

"Wilkinson makes the case that the evolutionary forces of individual selection and kin selection, operating simultaneously over eons, have produced the best and worst aspects of human nature. God's use of evolution in the creation of humanity therefore sets the stage for life to truly test our willingness to choose good over evil as we respond to competing urges. Well-researched, insightful, and provocative."

—Laura C. Bridgewater, professor of microbiology and molecular biology, Brigham Young University

"Into the midst of the often contentious debate about evolution and the meaning of life, *Purpose* breathes a breath of fresh air. By taking seriously the notion that 'everything is evolved,' Wilkinson highlights how evolution must then be responsible not just for our genetic material, but also for the apparent dual nature of human nature—the tension between selfishness and altruism or between aggression and cooperation. These aspects of human nature suggest that, while the steps of evolution are random, the higher order principles that guide evolution have all the apparent hallmarks of having a clear purpose. I highly recommend this to those seeking a clear and hopeful perspective on how modern science can help us pursue a meaningful life."

—Troy Van Voorhis, professor, MIT, and author of *Certainty: Is Science All You Need?*

"Dr. Wilkinson has given us a wonderfully—indeed masterfully—synthetic work on the biggest question of all. Bringing together insights drawn from the fields of biology, psychology, sociology, philosophy, and theology, he puts a spotlight on the things that give human beings, over time and across cultures, a sense of purpose. His book is an intelligent person's guide to the meaning of life."

—Robert P. George, McCormick Professor of Jurisprudence, Princeton University

"Anyone who wants to know how a scientific understanding of reality and the notion of purpose for human existence can go together, will be richly inspired by this book."

—Dirk Evers, professor of theology, Martin Luther University Halle-Wittenberg, Germany

"Wilkinson has done something extraordinary: He has provided a science-based answer to the world's most intractable philosophical question. And he has done it in elegant, entertaining prose that any thoughtful person can enjoy. This book will change your life. It tells us not just how to live, but why. It is especially inspiring to young people struggling to find purpose in an uncertain age. This book is for anyone who wonders about the meaning of life."

—John Morley, professor, Yale University

"In *Purpose*, Wilkinson deftly explains how and why our most fundamental institutions—including marriage and family—play a crucial role in grounding and guiding our lives. This is a fascinating and important book."

—W. Bradford Wilcox, sociology professor and director of the National Marriage Project, University of Virginia

"Wilkinson makes a powerful argument for how human nature, with all its complexity, suggests there is a purpose to our existence. Drawing upon the profound insights of evolutionary biology, this successful effort to reconcile sometimes competing worldviews gives a reasoned explanation that our existence is not accidental. Wilkinson demonstrates that what we have learned so recently about human nature from our evolutionary history confirms what we have known so long from the humanities: that the boundary between good and evil runs through every heart. That Wilkinson makes his case in a way that is accessible to the layperson will broaden and deepen the impact of this important contribution to understanding who we are and what we should be about."

—Thomas B. Griffith, former federal judge of the United States Court of Appeals for the District of Columbia

"In this lively, refreshing, and well-written book, Samuel Wilkinson thoughtfully explores the fascinating problems of evolution, freedom, meaning, and religion. Complex ideas are explained in simple and clear terms, and arguments on all sides are carefully scrutinized. Readers will enjoy engaging with this intelligent and humble mind. Anyone interested in the deep questions about human life will find this book a valuable and stimulating read."

—Roy Baumeister, president-elect, International Positive Psychology Association, and co-author of the *New York Times* bestseller *Willpower: Rediscovering the Greatest Human Strength*

"Samuel Wilkinson brings his considerable expertise to questions of evolution, purpose and God. What results is an innovative approach which takes the science seriously both in what it can say and where it needs a wider context to give insights into what it means to be fully human."

—Reverend Professor David Wilkinson, principal of St John's College, Durham University

"Teleology—explanations that privilege goals over causes—has long been eschewed in biology. Samuel Wilkinson challenges conventional wisdom with *Purpose*, his passionately argued book that seeks the essence of our existence by envisioning direction in evolution, not only randomness, and virtues in human nature, not only vices. Wilkinson, a psychiatrist specializing in depression, finds purpose in *Purpose*: to restore faith and bring meaning to many. An unabashed proponent of theistic evolution, Wilkinson argues that a purpose of our existence is to choose between the good and evil inherent within us—life is a test, he says—and he offers a framework to help us choose our better natures, maximize our individual well-being, and thus live the Good Life. While some (including me) may not entirely agree, all should at least consent that this debate sits at the heart of the human condition."

—Robert Lawrence Kuhn, creator and host of the PBS series *Closer to Truth*

PURPOSE

What Evolution and Human Nature Imply About the Meaning of Our Existence

SAMUEL T. WILKINSON

PEGASUS BOOKS
NEW YORK LONDON

PURPOSE

Pegasus Books, Ltd.
148 West 37th Street, 13th Floor
New York, NY 10018

First Pegasus Books cloth edition March 2024

Interior design by Maria Fernandez

Library of Congress Cataloging-in-Publication Data is available.

ISBN: 978-1-63936-517-3

10 9 8 7 6 5 4 3 2 1

Printed in the United States of America
Distributed by Simon & Schuster
www.pegasusbooks.com

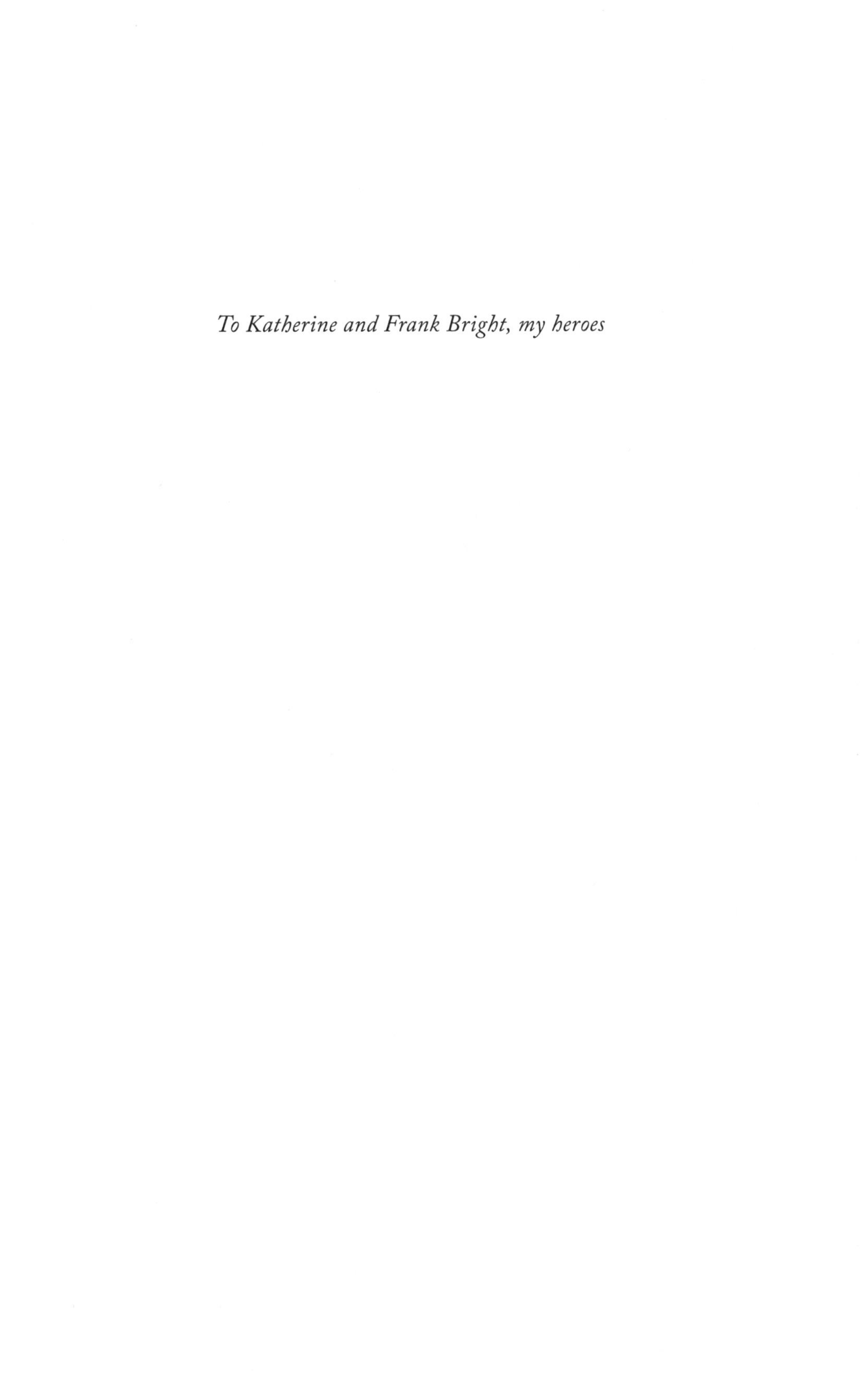

To Katherine and Frank Bright, my heroes

Contents

Preface

When I was in my first year of medical school, I became terribly vexed by the apparent contradiction between science and religion. I had attended an undergraduate institution where religious faith was encouraged. It was fairly easy to believe in this environment. But within three months of graduating college, I married my wife, and we moved to Baltimore, Maryland, where I enrolled in medical school at Johns Hopkins. I had some incredible experiences and made many wonderful friends at Hopkins. But, unlike my undergraduate institution, the academic environment of medical school was not one that nurtured religious faith. At the core of my doubts was the theory of evolution. I struggled to reconcile my belief in a benevolent God who created us with the seemingly mindless process of evolution. It would seem that if we are the products of a random process of evolution, then life would have no purpose or meaning.

After wrestling with this issue for many months, I experienced a sort of epiphany that shifted my understanding. Principles from many different disciplines seemed to suddenly come together in my mind in a beautiful and harmonious way. Evolution wasn't a totally random process; it had a direction and was guided by natural principles.

Furthermore, the way that evolution shaped human nature produced the strongest forms of love and affection within family relationships. Parents have a deep capacity to love their children because they evolved to do so. This is how God made us. This book is my best effort to explain how, as a result of my experience, I now view the world.

For me, this worldview has been immensely satisfying from both a scientific and spiritual perspective. I believe it also has the potential to help us answer, at least in part, some of the deep questions about ourselves. Is there a purpose to our existence? If so, what is it? Are we ultimately selfish or altruistic? What is at the root of human happiness? I hope that this book will provide some answers to those questions, and in doing so will help those who struggle to understand how current scientific theories of our origin fit with a belief in a God who created us and has a purpose for our existence.

In writing this book, I drew from a variety of disciplines, including evolutionary biology, psychology, philosophy, and sociology. Given that my narrow area of expertise focuses on the psychology of human depression and suicide, I have had to rely on the knowledge and enthusiasm of many different experts in these diverse areas. I borrow from their ideas heavily, with proper citation, in many instances quoting them directly. I am deeply grateful to many of these experts who generously reviewed excerpts from the book for technical accuracy.

Although one of my chief objectives here is to show how science and faith are not in conflict, I am mostly silent throughout the book about the particulars of faith. My reason for treading lightly in matters of theology and heavily in matters of science is that although most people believe in God, there is wide disagreement about the nature of God. But most religions are united in a few fundamental beliefs: the belief in a higher power, the belief in a purpose to our existence, and similar beliefs in practical teachings on how to live a moral life (the Golden

Rule, some form of sexual restraint, and an obligation to care for the poor, to name a few). When I invoke God in this book, I refer broadly to a Being who is benevolent, who created us, has a purpose for us, and wants us to be happy. Beyond that, I refrain from further speculation or discussion about other aspects of God or religion.

Chapter One
Science, Religion, and the Meaninglessness of Existence

Are God and Nature then at strife,
That Nature lends such evil dreams?

—Alfred, Lord Tennyson,
"In Memoriam A.H.H."

Nashville, Tennessee; January 1925

The vote was seventy-one in favor, five opposed. The Tennessee legislature had just passed a bill that made it a crime to teach any theory that contradicted the Bible. The new bill, known as the Butler Act, was intended to expressly prohibit the teaching of evolution in public schools. Two months after the state legislature approved the bill, Tennessee governor Austin Peay signed it into law. Several states—including Texas, Kentucky, and New York—had debated passing similar laws in preceding years. Some had even passed resolutions that *recommended* against teaching evolution in public schools, but Tennessee was the first to make it a crime.

The new law elicited intense controversy. Free speech advocates immediately began searching for a test subject—someone willing to intentionally break the new law and go to court to challenge the law. The American Civil Liberties Union ran an advertisement in a local newspaper: "Looking for a Tennessee teacher who is willing to accept our services in testing this law in the courts. Our lawyers think a friendly test case can be arranged without costing a teacher his or her job . . . All we need now is a willing client."[1]

John Scopes, a twenty-four-year-old football coach and substitute teacher in the small town of Dayton, Tennessee, volunteered. Less than a month later, Scopes was indicted by a grand jury for violating Tennessee's new anti-evolution law. (One of the many ironies of the whole ordeal was that Scopes privately expressed considerable uncertainty as to whether he had ever actually taught evolution!) His indictment kicked off what some would later call the Trial of the Century.

The so-called Scopes Monkey Trial turned out to be as much publicity stunt as formal legal proceeding. In fact, Scopes's idea of "volunteering" as a test subject did not originate with himself; rather, local town leaders had urged Scopes to step forward in hopes that the trial's publicity would help jump start Dayton's economy. The strategy seemed to work (if only for a few weeks) and the trial attracted big names. William Jennings Bryan, former three-time presidential candidate for the Democratic Party and outspoken apologist of religious fundamentalism, was called in to help prosecute John Scopes. Clarence Darrow, arguably the most prominent criminal defense attorney in the nation at the time, volunteered to help the defense.

As the trial approached in the summer of 1925, Dayton took on a full-fledged carnival atmosphere. Six blocks of the small town (whose population at the time was under 2,000) were converted into a pedestrian mall.[2] Merchants sold food and memorabilia, including Bibles and toy monkeys. There was even an exhibit featuring two chimpanzees, one of whom paraded about dressed in a plaid suit.[3] Also featured was a "missing link" between humans and chimpanzees—in actuality a fifty-one-year-old Vermont man with a receding forehead and protruding jaw who was unusually short.[4] Prominent journalists had come from all over the country to the small town of Dayton, Tennessee, to report on the proceedings. For over a week, the Scopes Monkey Trial dominated the front pages of the nation's major newspapers.[5]

The trial began in earnest on July 10, 1925. One young high school student, when questioned under oath, reportedly declared: "I believe in part of evolution, but I don't believe about the monkey business."[6]

As an initial strategy, the defense urged the presiding judge to strike down the anti-evolution law as unconstitutional. In an impassioned speech, defense attorney Darrow declared: "We find today as brazen and as bold an attempt to destroy learning as was ever made in the Middle Ages!"[7] When the judge declined to rule the law unconstitutional, the next major conflict centered on whether the defense would be permitted to call scientific experts as witnesses to support the theory of evolution. After several days of deliberation, the judge again ruled against the defense and barred the calling of scientific expert witnesses. This ruling by the judge essentially killed the core strategy of the legal defense.[8]

Finally, on the last day before the trial concluded, Darrow made the highly unusual move of calling Bryan himself, one of the key prosecutors, to testify as a Biblical expert. Bryan was happy to oblige the defense, confident about his ability to defend his faith, even under oath. His co-prosecutors, recognizing that the case of law had already been decided, urged him not to testify. But Bryan was insistent. He understood that the trial was about much more than factually establishing whether Scopes taught evolution in a public school and transgressed the Butler Act. Bryan viewed the trial—as did many—as a broader argument of science versus religion. At stake was not just whether the Butler Act would be upheld, but how people viewed the legitimacy of both science and religion—and how much attention these competing worldviews commanded in the public sphere.

The examination of Bryan by Darrow resulted in a painful and cruel legal renunciation of Biblical literalism. From a purely intellectual standpoint, Darrow clearly bested Bryan. But he did so in a way that was mean-spirited, and the immediate media response suggested that most of the public sympathized with Bryan.[9]

After eleven days of tense courtroom proceedings and sweltering heat (the judge allowed part of the trial to proceed outside the courthouse, partly as a respite from the weather[10]), a guilty verdict was returned. The jurors deliberated for only nine minutes (most of this time was used simply to leave and reenter the crowded courtroom).[11] Scopes was convicted of breaking the new anti-evolution law and was fined $100. Bryan and the prosecution had won the trial but were humiliated in the process.

Within a week of the trial's conclusion, William Jennings Bryan unexpectedly died in his sleep. H. L. Mencken, writing for the *Baltimore Sun*, quipped: "God aimed at Darrow, missed, and hit Bryan instead."[12] On his grave, it was inscribed: HE KEPT THE FAITH.

◆

The Scopes Monkey Trial, perhaps more than any other point in the modern era, has come to epitomize the debate between science and religion. The only possible exception was when evolution was originally described as a serious scientific theory by Charles Darwin himself. Darwin first published his work explaining the basic principles of evolution in 1859 under the title *On the Origin of Species*. Although he was mostly silent in this initial work regarding the origin of man, many of his colleagues immediately began to propose that such a theory obliterated longstanding views that man was the result of a special, supernatural act of creation by a Divine Being. (Just prior to the publication of *On the Origin of Species*, Thomas H. Huxley—later dubbed, affectionately, "Darwin's Bulldog"—wrote to Darwin declaring his willingness to defend evolution against fundamentalism: "I am sharpening up my claws and beak in readiness."[13]) In many ways, Darwin's theory of evolution has since become a divisive wedge between scientific and religious thought. As seen in the Scopes Monkey Trial, many since Darwin have refused to believe or objected to the teaching that evolution had anything to do with the origin of humanity. On the other hand, some outspoken atheists have preached that the theory of evolution explains away the existence of a Divine Creator; mankind, they assert, would be better off if we gave up our religious traditions. The conflict remains unresolved to this day.

Many have compartmentalized how we view the world into two main categories: a scientific world view and a religious one. When we ask ourselves how and why things behave the way they do, we invariably turn to science. When we search for answers to the deep questions of the soul—*Why does suffering exist? What is our purpose or meaning? Does life after death exist?*—many of us find that science as yet provides no satisfying answers. It is then that we turn to the religious traditions in an effort to appease the yearning to answer these deep questions.

There have been many attempts to reconcile these two great forces—science and religion—but in my opinion, there is still much work to be done in this area. The chapters that follow are my attempt to add to those ideas[14] that reconcile scientific inquiry with a belief in a Divine Creator who is somehow responsible for the order we see around us.

One of the main theses of the book is that evolution was the mechanism by which a Higher Power created all life, including humans. Most people would refer to this being as God (though a small portion deny the existence of God but still believe in a Higher Power of some sort). Most believe that God is benevolent and loves everyone despite their faults. Many believe that God has a higher purpose for our lives. When I invoke God in this book, I refer broadly to a Being who is benevolent, who created us, has a purpose for our existence, and wants us to be happy. Beyond that, I refrain from further discussion about other aspects of God's nature.

A large part of the reconciliation I am attempting focuses on resolving two central dilemmas with which I once struggled. Through intense study and reflection, I came to realize that these dilemmas do not pose the challenge to orthodox religion that I once assumed they did. This book stems from what I learned during my journey to resolve these dilemmas, which may perhaps help those who still struggle to reconcile evolution with a religious worldview.

Dilemma #1: The Doctrine of Randomness

The first dilemma has to do with what might be called the *doctrine of randomness*. If evolution, as is commonly misunderstood, was a random and haphazard process, then it would follow that human beings are merely intricate molecular accidents. This view of evolution seemingly obliterates any sense of universal meaning or purpose in life. As the celebrated Harvard biologist Edward O. Wilson declared: "No species, ours included, possesses a purpose beyond the imperatives created by its genetic history. Species may have vast potential for material and mental progress but they lack any immanent purpose or guidance from agents beyond their immediate environment or even an evolutionary goal toward which their molecular architecture automatically steers them."[15]

As I'll explain in subsequent chapters, I believe Wilson is wrong in his logic. Furthermore, Wilson's views have helped fuel the idea that we find ourselves alone in an indifferent universe, void of any absolute meaning or purpose.

From a traditional scientific standpoint, Darwin's theory of evolution has been agnostic toward or even antithetical to an intentional purpose of existence. If Darwinian natural selection was the principal operating force behind evolution, then genetic chance and environmental necessity are responsible for our existence. A vocal minority within the scientific realm—sometimes dubbed ultra-Darwinists—seem to have reveled in the obliteration of any higher purpose of existence by the supposedly random process of evolution.[16] This outlook is described by one scientist who declares with biting bluntness:

> In the purely natural and inevitable march of evolution, life . . . is of profound unimportance . . . a mere eddy in the primeval slime.[17]

As I've worked with hundreds of depressed patients over the years, I've found that many people struggle with a sense of purpose or meaning in their lives.* I find this tragically ironic. Compared to those of earlier time periods, people who live in developed countries today have so much to live for. They have greater access to medicines that can reduce the risk of death and disease. Through the marvels of modern

---

* There are often many factors that contribute to depression, but one that is so often overlooked by contemporary experts in mental health is a lack of purpose or meaning.

technology and the vast amount of information that is available, they can readily learn about the rich cultures of countless societies both past and present. Their opportunities for both education and leisure far surpass those of past generations. And yet so many struggle to find hope and meaning in their lives.

Across the globe, depression is on the rise and is now the leading cause of disability worldwide.[18] In the United States, so-called "deaths of despair" (which include suicide, unintentional drug/alcohol poisoning, and alcohol-related liver disease) have increased dramatically in recent decades.[19] The rate of suicide alone has increased by almost 40 percent in the last twenty years.[20]

These tragic trends can be seen not only at the individual level, but also among institutions. As I write this book, in the United States and many other countries, there is a growing distrust of organizations, of political leadership, a growing worry of extremism, a waning desire to belong to organized faith groups (or organizations of any kind). Certainly, there are many potential reasons for the growing cynicism in the modern world. But in my opinion, at the heart of it all lies a loss of faith. A loss of faith in a benevolent God. A loss of faith in the goodness of humanity. A loss of faith in an absolute purpose and meaning to our existence.

Not all of this growing tide of cynicism that increasingly pervades our culture can be traced back to Darwinism. But some of it can. Many seem resigned to believe that if evolution is true and is the natural process that created all life (including human beings), then we are doomed to live, breathe, die, and whittle away our hours in a world without meaning. In other words, we only exist because our genes—those clever "replicators"[21] that are miraculously condensed within each cell[22]—are good at reproducing and surviving. Hence, any notion of a higher purpose for our existence is likely to be met, from a strictly orthodox Darwinian perspective, as heresy.

Yet so many find this worldview unsatisfying, or at best, incomplete. People everywhere, rich and poor, educated and ignorant, yearn for a deeper meaning and purpose to life, a higher fulfillment of individual and collective potential. Lewis Thomas, a noted physician who served as the dean of the Yale School of Medicine, expressed this sentiment well:

> I cannot make my peace with the randomness doctrine; I cannot abide the notion of purposelessness and blind chance in nature. And yet I do not know what to put in its place for the quieting of my mind. . . . We talk—some of us, anyway—about the absurdity of the human situation, but we do this because we do not know how we fit in, or what we are for.[23]

Even among the most ardent supporters of evolution there are those who question this conclusion. If biological life is merely the result of the blind forces of nature, then why are we—as products of a random biological process—so driven to find meaning and purpose?

To overcome this first dilemma of the randomness doctrine, I endeavor to show 1) that evolution was not random but that it had (and may still have) a direction; and 2) how a deep understanding of human nature and the evolutionary heritage that shaped it implies that life does indeed have a purpose. In fact, through a framework that brings together principles of evolutionary biology, psychology, sociology, and philosophy, we can infer not only that there is a purpose to our existence, but *what this purpose is.*

Dilemma #2: Our Selfish Core

The second dilemma that we must overcome is what (in so many people's minds) evolution implies about human nature: that at our core we are greedy, aggressive, and self-serving. Given the guiding principle of natural selection (*survival of the fittest*), it's easy to understand why people would assume that the theory of evolution emphasizes the darker propensities of humanity. It became advantageous for the developing human (or hominin) to hoard resources during times of famine. If he wasn't selfish enough, he risked starvation and death, while a more selfish hominin survived and propagated the trait of selfishness. Aggression can be treated in a similar manner; at any threat of attack, the evolving hominin who was too weak or passive to defend himself perished, while a more aggressive peer survived. Unrestrained sexuality, of course, is all too easy to explain by virtue of the Darwinian theory of evolution. A prehistoric male hominin that was able to attract and mate with several females was at a marked advantage in propagating his genes when competing against his peer who exhibited a more restrained approach to mating. And so we could continue with many of the negative dispositions of human nature: greed, jealousy, disloyalty, arrogance, thievery, and aggression all have their origins in the operation of natural selection. An account in the late 1800s, now thought to be apocryphal but wonderfully expressive of the sentiment felt by many, tells of a woman who, upon first hearing of the theory of evolution and its implications, exclaimed: "My dear, let us hope it is not true; but, if it is true, let us hope it will not become widely known!"

In the minds of some, Darwin's theory of evolution created a bleak picture of a "nature, red in tooth and claw."[24] If we are the products of evolution, then it might imply we are selfish, aggressive, and greedy. William Jennings Bryan voiced this concern thus: "There is another objection [to evolution]. The Darwinian theory represents man as reaching his present perfection by the operation of the law of hate—the merciless law by which the strong crowd out and kill off the weak."[25] Does evolution, then, resign us to such a pessimistic view of human nature? Is it true that humans are nasty, brutish, and selfish? Even in the modern era, some of Darwin's fiercest defenders assert that our biological history leaves us in no condition to be moral creatures. Says evolutionary biologist Richard Dawkins: "Be warned that if you wish . . . to build a society in which individuals cooperate generously and unselfishly towards a common good, you can expect little help from biological nature . . . because we are born selfish."[26] For a lot of us, this is a bitter pill to swallow and a very depressing view of human nature. In the minds of many, a full acceptance that evolution is responsible for our natures means that we must surrender to the view that humans are ultimately self-serving creatures and all the terrifying implications that follow.

Yet over time, scientists have realized that the picture of human nature is complex. For example, even before they are able to communicate verbally, infants show a remarkable tendency to help others in need, even an adult who is a stranger. Dozens of experiments show not only that humans are remarkably altruistic, but that we have evolved in such a way that we find altruism rewarding.[27] In an experiment with children as young as twenty-two months, those who gave away candy were happier than those who received candy.[28] But where does this propensity for altruism come from? How does evolution create beings who are prone to give, even to the point of self-sacrifice in extreme circumstances? In short, there are several evolutionary mechanisms that have shaped human nature. In many instances, these evolutionary forces have pushed and pulled in different directions, thus creating opposing tendencies within us. While selfishness is indeed an unfortunate part of human nature, so too altruism is a powerful and innate capacity within each of us.

In later chapters, I hope to show in convincing fashion that evolution is not only responsible for the darker propensities of human nature, but it is also at the root of our prosocial and moral behaviors as well. This would then overcome the second dilemma and show that we are not ultimately selfish, at least not exclusively so. At our core, we are conflicted and pulled between selfishness and altruism. No wonder life is so fraught with tension and difficulty.

My overall thesis is that evolution was the means by which a Divine Being created all life, including us. If this is the case, then a careful study and proper understanding of evolution and human nature may give us some clues as to the purpose of our existence. In other words, *how* we were created may help us understand *why* we were created. Furthermore, a proper understanding of how we were created may also help us better comprehend human nature, how we can be happier, and how we can form societies that bring out the better angels of our nature.

Principle #1: Evolution was not random. Ever since Darwin, the default assumption has been that evolution was a totally random and haphazard process. In other words, if you wound back the clock several billion years and let evolution rerun, nature would have produced a totally different result. But recent discoveries have prompted many scientists to question this assumption. One of the biggest reasons for this is a biological phenomenon called *convergent evolution*. Convergent evolution occurs when species that are not closely related independently evolve the same structure or ability to solve a similar problem (like the problem of how to fly). For example, bats, birds, and butterflies all have wings and the capacity for flight. But their common ancestor had neither wings nor the capacity for flight. Bats, birds, and butterflies all *independently* developed the ability to fly. Echolocation, the ability to navigate space by sending out vocal signals and interpreting their patterns as they echo off physical structures, evolved independently in bats, dolphins, and a type of shrew. It turns out, virtually every biological trait or structure has evolved more than once. It seems that there have been higher-order principles that have given a direction to the process of evolution.

To be clear, I am not arguing that evolution didn't happen. But given the remarkable complexity of life, the assumption that the process was totally random must be questioned. Even Darwin's staunchest defenders have now acknowledged this. That some randomness was present is very likely. But there seem to be higher-order chemical and physical laws that have guided evolution to produce the same biological forms over and over again. Some scientists are even starting to speak of "deeper organizational principles that underpin evolution,"[29] recognizing that although the *course* of evolution may be random, the *outcomes* are not. This seems to be the case even if we don't understand all of these higher-order principles. Given the natural laws by which the universe is governed, life as we know it seems to have been inevitable.

Principle #2: In regard to the evolution of humans, nature seems to have created within us competing dispositions. As discussed above, what evolution implies about human nature is one of the key reasons that it has been (and still is, to many) so unsavory and difficult to accept. It's easy to imagine a narrative in which only the selfish, aggressive, and cruel survive. Yet the picture is complex. Several evolutionary mechanisms—including reciprocal altruism, indirect reciprocity, kin selection, individual and perhaps even group selection (more on these later)—have pushed and pulled us in different directions. As a result, selfishness, aggression, and cruelty are unfortunate parts of human nature. But so too are altruism, cooperation, and kindness. This is what I will call the *dual potential of human nature*. It should be noted that, with regard to our prosocial tendencies such as love, altruism, cooperation, and so forth, such dispositions likely became strongest in relation to those with which the evolving hominin most closely shared its genetic material. In other words, the positive dispositions were most strongly reinforced among members of the primitive family unit.

Principle #3: Free will is a key aspect of human nature. Despite significant advances in our understanding of neuroscience and human nature, free will remains elusive as a subject of scientific inquiry. Yet from a subjective perspective, the sense that we have free will is indisputable. Life is willful. No one thinks of him or herself as the passive follower of the brain or body. We each consider ourselves, at least to some degree, "in charge" and free to direct and control our attention, thought, speech, and action. Despite arguments against the existence of free will from well-meaning scientists and philosophers, a large and growing body of evidence supports the notion that we have free will. How evolution has endowed us with free will (or even how it develops from infancy to adulthood in a given individual) is still a great mystery. Furthermore, our possession of free will does not mean that there are not biological or environmental limitations to our freedom, or that culture does not have great influence over our decisions. But in the end, my interpretation of the psychological evidence is that, on some basic level, humans are free and responsible agents. This principle is critical to an accurate understanding of human nature and is central to the purpose of our existence.

As we consider the principles from above, it would seem that the purpose of our existence (or at least one such purpose) is to choose between these competing natures. To choose between altruism and selfishness, cooperation and aggression, love and lust. In language that has largely been discarded, a fundamental purpose of our existence is to choose between the good and evil inherent within us. Life is a test.

In the second half of the book, I build on these principles to reveal a framework for human nature that will help us learn how to choose our better natures (and thus establish a *Good Society*) as well as maximize our individual happiness and well-being (and thus live the *Good Life*).

Principle #4: Strong family relationships are key to the *Good Life*. Much has been written about the key to human happiness. While there is still much to be learned, it is now overwhelmingly clear from a huge amount of psychological evidence that *relationships with others* are the most important factor for our happiness and satisfaction.[30] I argue that this must have come from evolution. In past eras when resources were scarce and disasters were frequent and unpredictable, it paid off to have strong relationships and a close-knit group. So much so that forming and maintaining deep relationships became a critical part of our emotional health and, consequently, became deeply rewarding. It would therefore stand to reason that the relationships that are most relevant to evolution (and most central to our happiness) are those with whom we share genetic material: our families. As a brief example, consider human offspring. Because our newborns are so fragile when they are born, they evolved to elicit a deep emotional response on the part of their parents to love and care for them, otherwise they would not have survived. And indeed, this is the case. Any parent can attest to the deep feelings of love that are awakened as you first hold your newborn child. This is the way evolution has shaped us. Nothing in our lives will be more rewarding than forming and maintaining high-quality relationships with our family members. This is the way we are evolutionarily engineered. This is how God made us.

Principle #5: Strong family relationships are key to the *Good Society*. Given the *dual potential of human nature*, a wealth of data demonstrates that strengthening family relationships can help people choose their better natures.[31] This is especially true for men (who by and large are predominantly responsible for violence, crime, murder, abuse, and other social ills). When men are actively engaged in their social roles as fathers, their better nature tends to predominate. A key overriding conclusion of sociology is that engaged parenting helps us to become less selfish and to be law-abiding citizens. As one writer describes, "Fatherhood, more than any other male activity, helps men to become good men: more likely to obey the law, to be good citizens, and to think about the needs of others. Put more abstractly, fatherhood bends maleness—in particular, male aggression—towards prosocial purposes."[32] This is because the strongest forms of altruism, love, and cooperation come from the way that evolution has shaped our family relationships. Being involved in the providing for and rearing of our own children is nature's most powerful way to nudge us to choose our better natures. So many of our most urgent social problems can be tied to the dissolution of family relationships, especially between men and their children. In short, the key to the *Good Life* and the *Good Society* is, as much as possible, to strengthen relationships between children and their parents. This is a result of the way human evolution has shaped our natures.

◆

A story is told of a man who is repairing his roof. Suddenly, he loses his balance and starts to fall. He utters a quick prayer, "God, please help me and I will repent of my sins!" Just as he concludes his prayer, a roofing nail catches his clothing and stops his downward momentum.

"Oh, never mind, God. My pants got caught on this nail." This story has an important point: so often we think that if God works, it must be in ways that we don't understand. After all, tradition holds that God works in mysterious ways. But I don't think it was intended that God's ways remain mysterious forever. The man falling off the roof was saved by a nail; but who's to say that God didn't somehow use the nail to save the man? Eventually, I believe we can learn the ways of God, including the means by which we were created. I believe these means include some of the principles of evolution. In a famous essay in 1973, evolutionary biologist Theodosius Dobzhansky declared: "It is wrong to hold creation and evolution as mutually exclusive alternatives. I am a creationist and an evolutionist. Evolution is God's, or Nature's, method of Creation."[33]

Although one of my chief objectives in this book is to show how science and faith are not in conflict, I am mostly silent about the particulars of faith. But just because I tread lightly in matters of theology, and heavily in matters of science, doesn't mean I don't think some divine force is behind it all. For instance, throughout the book, I often refer to "nature" or "evolution" as a force that has shaped, molded, and affected our psychological capacities. If you'd like, you can substitute the word "God" into these passages. To me, it's much the same. As I intend to show in Chapter 2, evolution wasn't random. It was God's mechanism for creation.

For people of science who have spent their lives creating a framework based on natural laws, it is sometimes difficult to accept the supernatural God described in the Judeo-Christian tradition. This God's miracles seem to defy the laws of nature. But I believe this is a false dichotomy. In fact, in my view, part of God's power comes from a complete knowledge of all the laws of nature.[34] God works in accordance with the laws of nature. However, sometimes we mortals become so arrogant as to forget that we do not understand *all* the laws of nature (not by a long shot). I think this misunderstanding about the relationship between God and nature is part of the supposed conflict between science and faith. I agree with Nobel laureate Charles Townes, who said:

> Science tries to understand what our universe is like and how it works, including us humans. Religion is aimed at understanding the purpose and meaning of our universe, including our own lives. If the universe has a purpose or meaning, this must be reflected in its structure and functioning, and hence in science.[35]

My hope is that, by the end of the book, the reader can recognize not only that evolution and religion are not in conflict, but that the way evolution has shaped our nature reflects the purpose and meaning of our existence. Here goes.

Chapter Two

Evolution: Random Chance or Guided Process?

"Once is an instance. Twice may be an accident. But three times or more makes a pattern."[1]

Ever since Darwin, some biologists have preached that evolution was a completely random and haphazard process. This is one aspect of evolution that has been difficult to accept for some of us, especially those who believe in God. A process that is completely random would seem to be at odds with a God who intentionally created us and has a purpose for our existence. This is the dilemma of the doctrine of randomness. A traditional approach to evolution would suggest, in essence, that if you wound back the clock several billion years and let evolution rerun, nature would have produced a totally different result.[2] But recently, another view is emerging, one that recognizes that evolution is not totally random and unpredictable. While evolution may begin with random mutations, the outcomes of evolution seem to be guided by natural laws and constraints. The principal reason for this emerging nonrandom picture of evolution is a phenomenon known as *evolutionary convergence*. To be sure, the arguments in this chapter don't prove that there is a universal purpose to our existence; but they do provide quite a bit of room to allow for this possibility.

Convergence Rather Than Divergence

One of the linchpins of Darwin's theory is that all life descended from a common ancestor. As time slowly marched forward, species diverged from this common source, driven by different environmental and survival pressures. This has led to the wonderful diversity of life that we enjoy today. Depictions of an evolutionary tree of life, such as Figure 2.1, show how each large group of animals all descended from a common ancestor. For instance, mammals, reptiles, fish, and birds are all known as chordates. They are part of a large grouping of animals called the Chordata phylum. It is presumed that all mammals (including ourselves), reptiles, fish, and birds thus descended from a common, more primitive species. In turn, the chordates as well as other animal groups such as arthropods (crabs, lobsters, insects) and mollusks (octopuses, snails, squids) are part of the greater animal kingdom. Hence, all chordates, arthropods, and mollusks share an even older, more primitive ancestor. Finally, the animal kingdom along with other kingdoms (including plants, funguses, bacteria, slime molds, and other wonderful organisms) all descended from a common, single solitary cell.[3] This was the first cell that emerged from the primordial soup that scientists tell us existed some four billion years ago on Earth.

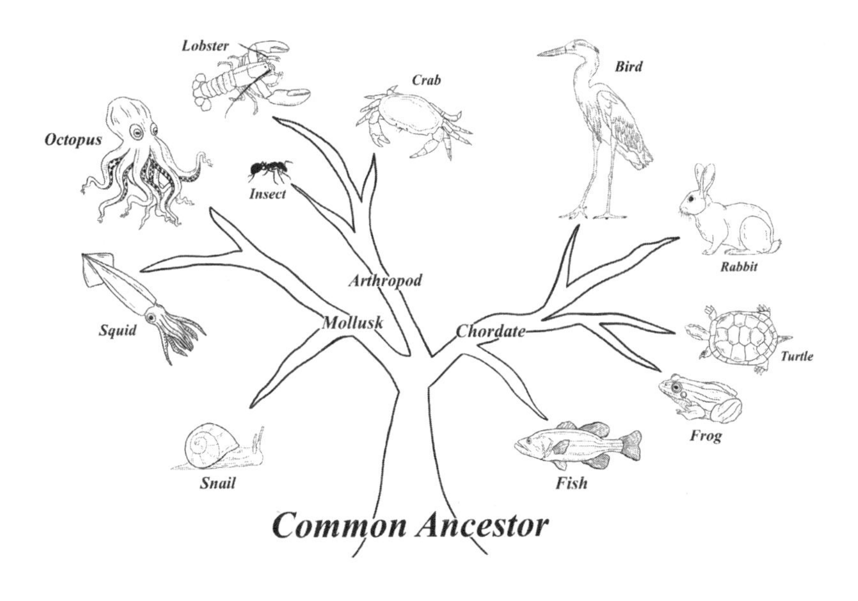

Figure 2.1. Depiction of an evolutionary tree.

As a given branch of the tree of life splits, species diverge. They develop different structures, fill different ecological roles, and go their separate evolutionary ways. But since Darwin, biologists have increasingly noted a somewhat curious finding. Even after species diverge, they will often develop similar structures to solve similar problems. In other words, despite different starting points on the tree of life, evolution will guide species to develop the same structures or functions.* This is called *convergent evolution.* Scientists once thought that examples of convergent evolution were unusual; however, nowadays it is recognized as being extremely common.

---

* Increasingly, even cellular and molecular processes (including genetic sequences) that developed independently have been discovered.

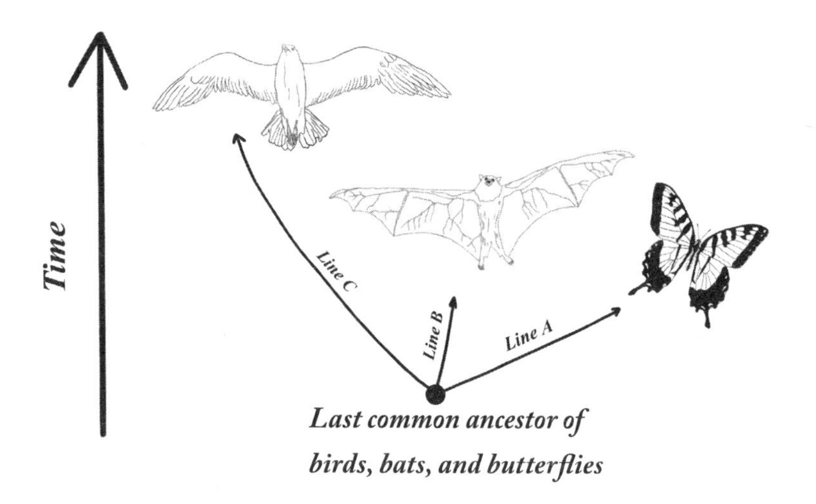

Figure 2.2. A graphical demonstration of the convergent evolution of wings. The last common ancestor of birds, bats, and butterflies did not have wings or the capacity for flight. Wings developed independently at least three separate times, along the lines labeled A, B, and C.

Consider the wings of birds, bats, and butterflies (Figure 2.2). These structures are quite similar in appearance and function. They are constructed in similar ways: a taut membrane tethered to a rigid frame. The common ancestor of these winged creatures (which lived some 550 million years ago) did not fly. Hence, birds, bats, and butterflies all developed wings and achieved flight independently. In other words, these creatures each evolved the ability to fly separately; they did not inherit this capacity from a common ancestor.[4]

Or consider the dolphin and the shark. The dolphin is a mammal with a skeleton of bone, and the shark is a fish with a skeleton of cartilage. Their last common ancestor did not have anything like the appearance of either today. Yet sharks and dolphins today look remarkably similar (Figure 2.3). Somehow, even though sharks never left the water and dolphins' ancestors were land-dwelling creatures, each independently evolved streamlined bodies and dorsal and pectoral fins to solve the problem of how to swim fast. Many types of sharks and dolphins are also shaded in such a way that they are dark gray on the top of their bodies with a light underbelly. This form of countershading makes them less visible from both above and below, which helps them sneak up on prey or avoid detection from predators.[5]

Figure 2.3. Convergent forms in dolphins and sharks.

Then there's echolocation, or the ability that some animals have to navigate space by sound. This remarkable faculty is another example of convergent evolution that occurred in bats,[6] toothed whales and dolphins,[7] birds,[8] and even some types of shrews.[9] These animals can send out vocal signals and then interpret their patterns as they echo off the physical structures of the local environment. Some animals are so good at this that they can detect objects as small as a mosquito or as thin as a human hair.

Think about the production of silken thread. Several insects, including spiders, certain types of moths, and weaver ants all produce it. It's a remarkable substance, as is the process of producing it. Spider silk, for example, starts out in a goopy, liquid form called dope. As the dope leaves the spider's body, its proteins must quickly reorganize themselves to transform into silk. The timing is almost impossibly precise: If the transformation happened at any point before leaving the spider's body, silk would clog the spider's glands; if it happened even just a tiny bit later than it does, silken thread would never form, and instead the spider would simply spew liquid.[10] And yet biologists tell us that some form of this exquisitely delicate process has evolved independently in at least twenty-three different types of animals.[11]

Perhaps the most remarkable and often-cited example of convergent evolution is that of the camera-type eye. This type of eye—which is the type that we have—consists of several elements: a transparent outer cover that can refract light, a lens with the ability to variably focus light, a diaphragm whose diameter adjusts with the level of brightness, and muscles that move in precise harmony with those of the other eye and adjust instantaneously to the fine tunings of the gyroscope of the inner ears. As the scientists Paul Bloom and Steven Pinker have written, "Structures that can do what the eye does are extremely low-probability arrangements of matter."[12] Yet the camera-type eye has evolved independently about half a dozen times (or so far as we have discovered, there may be more), including in a group of marine animals who are close relatives of earthworms.[13] The squid also has a camera-type eye that evolved independently and shares remarkable similarity to our own (Figure 2.4).

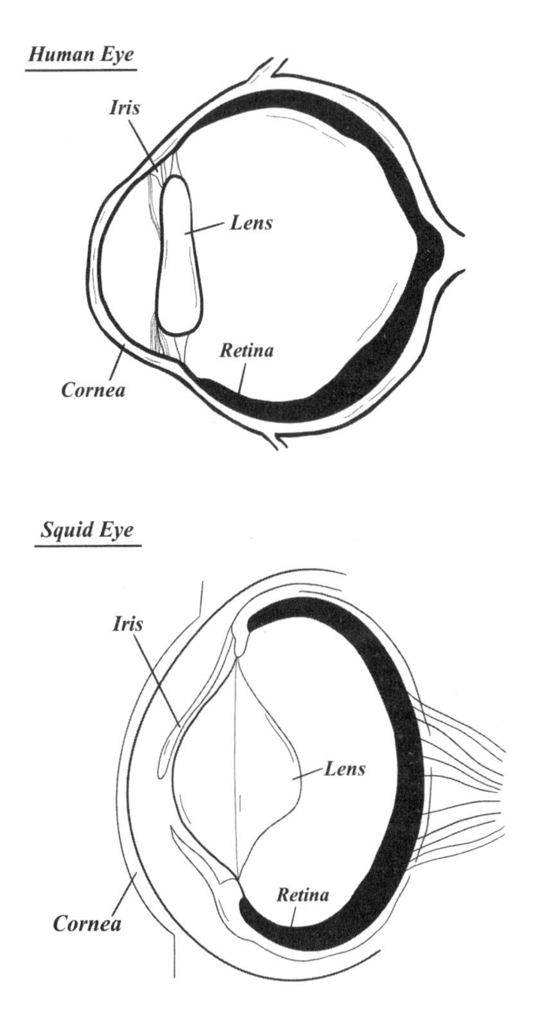

Figure 2.4. Convergent forms of eyes in humans and squids.

Scientists tell us that the last common ancestor of humans and squids was alive maybe 500 million years ago. But at that point, our species diverged. Yet we each developed camera-type eyes independently.

Of course, camera-type eyes are not the only type of eye that exists. Compound eyes have also developed independently in not-so-closely related organisms, among them insects and a type of clam. In contrast to a camera-type eye, compound eyes have multiple lenses that all develop almost on top of each other. From an evolutionary standpoint, compound eyes may be somewhat easier to develop than camera-type eyes, yet they are still extraordinarily complex.[14]

Altogether, experts estimate that eyes in some form or another have evolved independently at least forty times across different species.[15] Somehow, complex species that live where there is light will find a way to make eyes. As Richard Dawkins puts it, "It seems that life, at least as we know it on this planet, is almost indecently eager to evolve eyes. . . . If there is life on other planets around the universe, it is a good bet that there will also be eyes, based on the same range of optical principles as we know on this planet. There are only so many ways to make an eye, and life as we know it may well have found them all."[16]

Convergence can be seen at much smaller levels as well. There are many examples of convergent evolution seen inside the cell. For example, it can be seen with photosynthesis, the extraordinary process whereby plants extract energy from the sun and store it in the form of sugar. Today scientists estimate that a particular type of photosynthesis (called C4 photosynthesis) evolved independently at least sixty different times in different species of land plants.[17]

◆

Convergence: The Rule, not the Exception

Increasingly, biologists are realizing that in evolution, convergence is the rule, not the exception.[*18] In fact, as Cambridge paleobiologist Simon Conway Morris comments: "Convergence is ubiquitous. I can't think of anything which has only evolved once—or very, very few exceptions."[19] Richard Dawkins similarly notes, "It is surprisingly hard to think of 'good ideas' that have evolved only once." Indeed, Dawkins once made that point by challenging his colleague George McGavin to name some evolutionary adaptations that evolved only once—and McGavin could only think of a few.[20]

As we learn about more and more examples of convergence, we're beginning to develop the understanding that evolution is not a random process. "The visible results of natural selection and other shaping processes," writes Connie Barlow, "are by no means best characterized as

---

* In the above section, I document just a few instances of convergent evolution. Other scholars have made far longer and more comprehensive lists. For instance, the evolutionary biologist Simon Conway Morris, who is probably the world's leading expert in convergent evolution, has created a website dedicated to documenting examples of convergence. Here, Conway Morris lists over 1,500 examples of convergence, ranging from defense mechanisms (such as the ability to make oneself camouflage or secreting toxic or noxious chemicals [tetrodotoxin]) to biochemical processes (such as sodium voltage-gated ion channels, the convergence of several enzymatic processes, or the development of fungicide or herbicide resistance) to metabolic processes (such as respiration, hemoglobin [as well as myoglobin] in animals, or lysozyme enzymes). Other curious examples are the convergent development of venom and venom fangs in lizards, snakes and synapsids; luciferin (the group of enzymes that makes possible bioluminescence); carnivorous plants (at least six times); the defense mechanism of anointing (where animals give themselves the scent of poisonous animals so as to ward off potential predators) has occurred independently in rodents, hedgehogs, and tenrecs; the development of latex in plants and fungi; the development of immune systems in plants and animals; and the development of transparent tissues. See www.mapoflife.org for additional interesting examples.

a random pageant of form and function."[21] In exploring how the same skeletal forms occur over and over again, the biologists Roger Thomas and Wolf-Ernst Reif conclude, "While chance plays a preeminent role in setting the *course* of evolution, it contributes much less to the determination of its *outcome.*"[22]

Consider the drawing in Figure 2.5. Suppose you need to travel from point A to point B. There are four possible paths, and you have to roll a die to determine which one to take. This means that chance will determine the course you take—but no matter which one you take, the outcome will be the same: you'll arrive at point B. This is a little bit like the emerging view of evolution that convergence has given us in the last few decades. Some scientists even speak of "deeper organizational principles that underpin evolution,"[23] recognizing that although the course of evolution may be random, the end-forms are not.[*24]

So what accounts for the fact that convergence is so widespread, especially given the assumption that random genetic mutations underlie evolutionary change? One very real possibility (although it challenges traditional scientific thinking) is that the mutations were not all random.[25] This radical shift in one of the bedrock assumptions of Darwinism would not destroy the theory of evolution. Even Richard Dawkins, one of Darwin's staunchest modern defenders, acknowledges this: "It is not *necessary* that mutation should be random in order for

* Convergence has even been seen at the genetic level. In other words, the sequence of amino acids that makes up a given protein (as coded by the gene) can be convergent. In a way, this rocks the foundation of how the field of biology traces back and reconstructs the tree of life. If genetic convergence exists (which it does), and is increasingly seen as prevalent, then how do we know which creatures are actually related and descended from a more recent ancestor? The short answer is, it's impossible to know for certain given the limitations of our current framework of the scientific method.

natural selection to work. Selection can still do its work whether mutation is directed or not."[26]

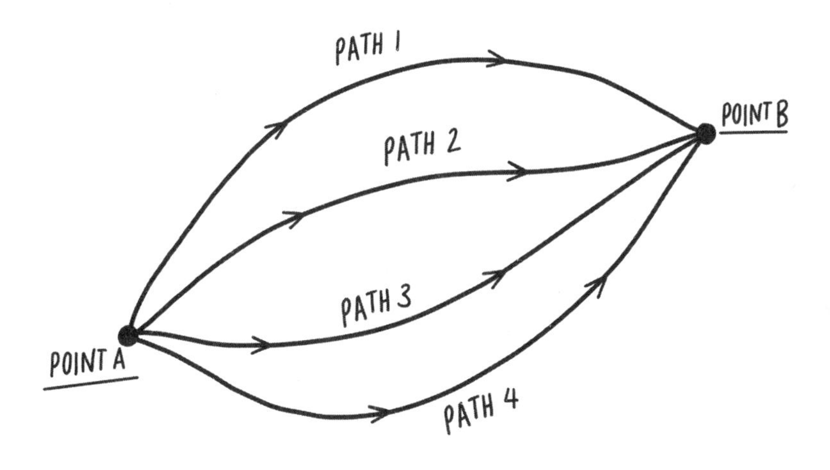

Figure 2.5. Four paths converging on the same endpoint.

But for the moment, let's put aside the possibility that the genetic mutations were somehow divinely directed (through natural laws and processes), and fall in line with mainstream science by assuming that the mutations were random. How is it possible for evolutionary convergence to be so widespread if the mutations are random? There's a lot we don't yet understand about this area, but as part of the explanation, it seems that evolution is greatly constrained. In other words, there are certain laws and principles that limit the number of possible outcomes that evolution can stumble upon. Natural selection is constrained to choose from a limited number of options. Hence, with a finite set of options from which to choose, if evolution continues choosing, it will invariably come up with the same solutions to life's challenges (i.e., how to find food, how to move around, how to sense the surrounding environment, how to avoid being eaten, and how to reproduce).

There are at least two principal ways in which evolution is limited in the number of options from which to draw upon. First, evolution simply has no use for many theoretical forms or structures; it "prefers" only those that confer some kind of advantage. Second, many constraints placed on evolution are related to the laws of physics and chemistry (for instance, the geometric and material properties that evolution has to work with). Even if a hypothetical structure might confer an advantage, there may be no way for nature's building blocks to combine to create it.[27]

The Museum of All Possible Animals

In describing the various life-forms that evolution has produced, Richard Dawkins imagines a Museum of All Possible Animals.[28] In this museum, animals are displayed next to each other in such a way that as we move along one corridor, a single structure varies ever so slightly. For instance, if we move east, the necks gradually become longer and longer. If we turn around and move west, the necks become shorter. Suppose we now choose to move in a different direction along another corridor. If we move north, the neck length remains the same, but something else changes, say the horns become longer. If we turn around and walk south, the horns become shorter. Finally, if we ascend the elevator to move to a higher floor, the neck and horns stay the same length, but the teeth become sharper. In constructing this imaginary museum, we quickly realize that there are more characteristics than just neck length, horn length, and sharpness of teeth. In fact, there are tens of thousands of characteristics in form that can vary ever so slightly. We thus come to the sobering realization that a space that is merely three dimensional is completely inadequate to describe this Museum of All Possible Animals. Instead, we need a space with tens of thousands (or even hundreds of thousands) of dimensions. At this point, the museum is becoming almost impossibly difficult to imagine (unless you are a mathematician or physicist). We also recognize that the museum is unfathomably large. In fact, the museum has far more exhibits in it than the number of atoms that exist in the known universe.

To take advantage of this idea but make it easier to imagine, let's think about a Museum of All Possible Shells.[29] Shells are hardened, calcified exoskeletons that provide protection and shelter for snails, clams, and other similar animals. Using a bit of oversimplification, and drawing on work from the 1960s by biologist David Raup,[30] we can construct a world of all possible shells that is represented by three factors based on how the shells develop: size, coiling, and elongation.[31] Out of all theoretically possible shells, however, only a very small number are actually found in nature. If so many shapes are possible in theory, why do only a relatively small portion of them actually exist? As discussed above, natural selection might have no need for them, or there might be no way that they could be created given the laws of nature.

The same dynamic would appear if one created a Museum of All Possible Skeletons. Roger Thomas and Wolf-Ernst Reif have imagined such a museum,[32] and in doing so they have found that although many skeletal systems are theoretically possible, only a few actually exist in nature. The shapes and designs of these skeletal forms, Thomas and Reif conclude, are "highly constrained by geometric rules, growth processes, and the properties of materials. . . . Given enough time and an extremely large number of evolutionary experiments," they write, "the discovery by organisms of 'good' designs—those that are viable and that can be constructed with available materials—was inevitable and in principle predictable."[33]

Now let's return to the Museum of All Possible Animals. As we've noted, the museum contains an extremely large number of exhibits, a number so large that it is virtually impossible for our minds to comprehend. Yet the proportion of animals that actually does exist is a tiny fraction of the total number of theoretical possibilities. And why is that? Because higher-order laws, principles, and constraints have directed evolutionary change (although our understanding of how this happens is still very limited). These "boundary conditions," as physicists and mathematicians might call them, all guide the process of mutation to form the biological world that we now enjoy and are a part of. When considered from this perspective, evolution is clearly not a random process. Even Richard Dawkins agrees. Evolution, he writes, is "*not* a theory of random chance. It is a theory of random mutation plus *nonrandom* cumulative natural selection."[34]

It may be counterintuitive that a process can be simultaneously random and guided. At the microscopic level, some evidence suggests that randomness is driving evolutionary change. But when we take a step back to the macro level, we get a sense that the process as a whole was not random.

It might be helpful to compare this process to a common manufacturing technique known as injection molding. This technique is one in which a solid material (commonly a type of plastic) is heated until it melts and becomes liquid. The hot liquid is then injected into a hollow mold of a predetermined structure. As the material cools, it regains its solid form. When the hollow mold is removed, the material retains the shape of the mold. This process is used to make many household and other common items, such as toothbrush handles, car parts, plastic bottles, and toys such as LEGO™ pieces (see Figure 2.6).

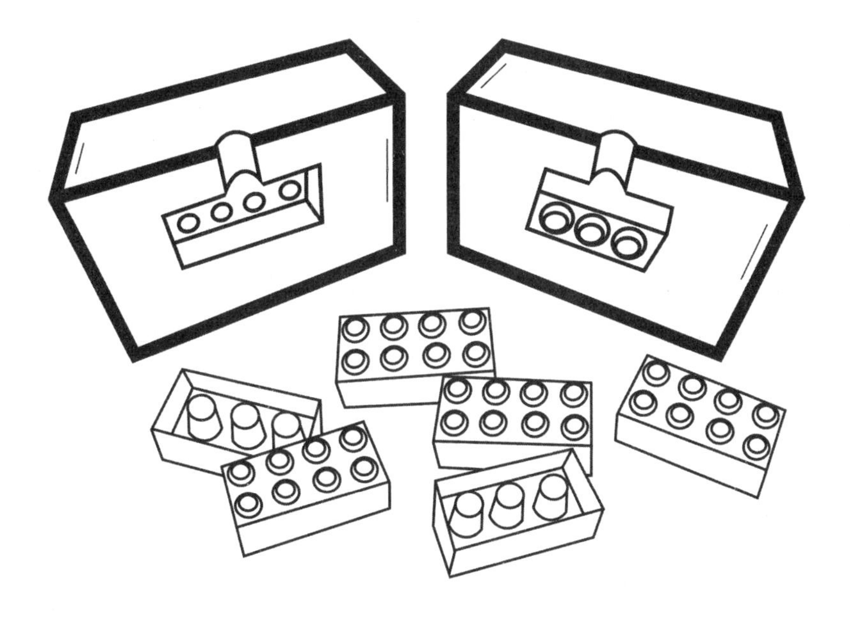

Figure 2.6. An injection mold and the LEGO pieces it formed.

If we observed the process of injection molding at the microscopic level, the movement of the molecules would indeed appear random. But when we take a step back, we see that the random motion of the molecules is constrained by the boundaries of the mold. At a macroscopic view, this is clearly not a random process.

In the case of evolution, imagine a mold with a cavity shaped like an intricate tree. As the liquid material fills the ecological void of the mold, it forms the evolutionary tree of life. The genetic mutations that drive the process may appear random from up close, but these movements are guided by the constraints of physical, chemical, and biological laws (in this analogy, the physical boundaries of the mold). These constraints guide the mutations to form the biological world that we are a part of. When we take a step back and view evolution from a bird's eye view, it is evident the process was not random.

Was Evolution Improbable or Inevitable?

Since Darwin, serious scientists and devout believers who are skeptical of evolution have struggled to comprehend how organisms as incredibly complex as humans, tigers, or even relatively simple bacteria could have evolved out of mere genetic chance. For instance, the physicist Fred Hoyle described being "constantly plagued by the thought that the number of ways in which even a single enzyme could be wrongly constructed was greater than the number of all the atoms in the universe, so try as I would, I couldn't convince myself that even the whole universe would be sufficient to find life by random processes—by what are called the blind forces of nature." Hoyle, who was no friend of religion,* went on to conclude: "A common sense interpretation of the facts suggests that a superintellect has monkeyed with physics, as well as with chemistry and biology, and that there are no blind forces worth speaking about in nature. The numbers one calculates from the facts seem to me so overwhelming as to put this conclusion almost beyond question."**[35]

Albert Szent-Györgyi, a biochemist who won the Nobel Prize, had similar thoughts:

* Hoyle was known to despise organized religion and resisted the Big Bang Theory so intensely (even though he was the one who gave the theory its catchy name) because he thought it implied a god-like quality to the genesis of the universe.

** Elsewhere, Hoyle compared the probability that evolution would randomly stumble upon the complexity of life we now observe to a tornado passing through a junkyard and randomly assembling an airplane, or 2,000 blind men simultaneously solving Rubik's cubes. Again, I think that Hoyle was right in this regard: that evolution is not the result of the "blind forces of nature." In other words, it was not random. It was constrained and guided by natural laws and higher-order principles.

> I have never been able to accept fully the idea that the adaptation and the harmonious building of . . . complex biological systems, involving simultaneous changes in thousands of genes, are the results of molecular accidents. . . . The probability that all of these genes should have changed together through random variation is practically zero, even considering that millions or billions of years may have been available for the changes. I have always been seeking some higher organizing principle that is leading the living system toward improvement and adaptation.[36]

I think both Hoyle and Szent-Györgyi were right to question the assumption that evolution is random. Randomness is *part* of the process, but it seems to have played a much smaller role than many originally assumed.

In trying to help both scientists and the general public understand how nature, through random and gradual means, has created the immense complexity and order of biological life, Richard Dawkins likens the situation to finding ourselves atop an almost unimaginably high peak named Mount Improbable.[37] From one vantage point, we can look down from where we are standing and see staggeringly high and sheer cliffs rising up from where we started. Could we really have climbed those cliffs? It seems impossible. But it's a matter of perspective, Dawkins argues. We need to shift our gaze to the other side of the mountain, where a long, gentle incline allows the traveler to gradually ascend with slow, methodical, sometimes imperceptibly short steps. What's missing from that analogy, however, is convergent evolution—the fact that, even on the gentle side of the mountain, we've made our way up countless times using the same pathways. Mount Improbable, in other words, seems actually to be Mount Inevitable.

At this point, we are beginning to get the sense that there really are some deeper forces and higher-order principles that guide organisms to independently develop similar structures and functions. When Darwin first proposed evolutionary theory, these constraints were not well recognized. Biology did not necessarily have higher-order principles, as there are in physics or chemistry. If you somehow traveled to another planet, a physicist who knew the mass and size of the planet could tell you the exact trajectory of a ball that you would throw. But it has always been assumed that a biologist would know nothing beforehand about any extraterrestrial life-forms that she encounters. This assumption, however, now seems to be wrong. Biology may indeed have higher-order principles and natural laws that govern the development of life-forms, even if we don't as yet fully understand them.

Ever since Darwin, biologists have assumed that evolution does not know where it is going. It is unguided and haphazard, analogous to a "blind watchmaker."[38] After all, mutations do seem to be random. But on a higher level, when we take a step back, biological evolution is constrained.* There are really only a handful of solutions to the common challenges that our world presents to living creatures (i.e., how to move, how to get food, how to sense the surrounding environment). There

---

* Another helpful way to understand the limited but potentially important role of randomness in evolution could be to consider a scientific experiment. In many scientific experiments, an element of randomness is intentionally introduced. This is the case in randomized clinical trials, where patients are randomly assigned to one of two or more intervention groups. If the experiment is carried out correctly and with enough participants, the randomness ensures that any difference between groups at the end of the experiment is only attributable to differences between the interventions that are being studied. In this way, randomness is a critical part of the experiment, but we would never describe the entire experiment as if it were carried out randomly. Indeed, any good scientist will conduct her experiments with utmost meticulousness and attention to controlling other variables. Hence, we call these *randomized* experiments, not *random* experiments. In the same way, randomness played a role in evolution (likely an important one), but the process in total was decidedly *not* random.

are, therefore, a limited number of arrangements of biological matter that evolution has achieved. Let me emphasize that in no way am I trying to belittle the wonderful complexity and diversity of life. With an estimated five million different species on Earth,[*39] it is impossible for the human mind to comprehend the breadth and depth of the biological world. But at the same time, compared to the number of structures that are hypothetically possible (which, by all means, is an unfathomably large quantity), the number of species that nature has achieved is only a minuscule fraction.

Embryonic Development: Evolution in Fast-Forward

To help solidify the concept that evolution was not random, consider the rather remarkable process of embryonic development, whereby a single cell made by the union of egg and sperm (an embryo) self-organizes into a full-fledged organism. In a sense, this could be considered a little bit like watching evolution in fast-forward. In humans, this remarkable process occurs over nine brief months. But the self-organization process that results in critical tissues and organs—lungs, kidneys, brain, and so forth—is actually even shorter, concluding within about ten weeks. Within three weeks, the developing embryo has a heart that starts to

* It turns out it's quite difficult to estimate how many total species exist on earth, and there is a lot of uncertainty in the estimates that do exist. On the one hand, a paper in the prestigious journal *Science* estimated there are around five million species on Earth. On the other hand, another analysis suggests there are around one trillion species. On the face of it, this is a huge amount of uncertainty. Imagine receiving a bank statement saying that your account has somewhere between five dollars and one million dollars! However, even with this great degree of uncertainty, in the world of orders of magnitude, the biodiversity that actually exists in nature is only a tiny fraction of the hypothetical forms that could exist.

beat. Approximately eight weeks later, it stops being called an embryo (derived from Latin and Greek, signifying "swollen") and is known as a fetus (derived from Latin, meaning "fruitful").[40] In a truly remarkable feat of nature, "it takes just forty-one cycles of cell divisions to get from conception to a fully formed little human."[41]

An embryo organizes itself in a most remarkable way. Every person reading this book, indeed every human that has ever lived, began their mortal existence as a single cell. How our intricate bodies develop from this humble origin is still quite a mystery. Of course, we know that the development of an embryo involves genes and proteins and cell replication and other complex biochemical processes. In embryonic development, there seems to be a certain flow or cascade of biochemical events, as genes are "turned on" in concert with other genes that are simultaneously "turned off." But modern biochemistry has not taken us much further than this. For instance, we know that a set of genes called *Hox* genes are extremely important in embryonic development;* they help determine the overall body plan, including which side of the embryo will become the head and which the tail, and how many arms and legs will be formed. But we don't understand most of the details behind how these *Hox* genes do this. No one has answered the deeper question of what makes self-organization possible. We can only assume that there are additional natural laws and principles that govern the remarkable process of self-organization as it pertains both to embryonic development as well as to the evolution of complex creatures. But as we consider and contemplate the process of embryonic development, we recognize immediately that it is most definitely not an accident, nor is it random. Nature knows what it is doing.

---

* In fruit flies, a single mutation at a specific *Hox* gene site can result in an extra pair of wings!

We are thus tempted to extrapolate this back to the process of evolution itself.[42] On the molecular level, the mutations may be random. But when we take a step back, it seems there are higher-order natural laws in play (many of which we have not yet discovered) that are guiding the process and have led to thousands, or even tens of thousands, of evolutionary convergences.

In sum, it seems that both random and nonrandom events played important roles in evolution. For example, the fact that you have blue eyes may be determined by a random reassortment of your parents' genes; but it is not random that you have eyes, and that you have camera-type eyes that can refract and focus light, with the almost identical structure that has evolved independently in squids and annelids. It seems that the development of eyes is written into the inherent self-organizing principles of biological matter. On a larger scale, it would seem that life as we know it was not an accident.[*43]

◆

To be sure, the widespread nature of convergent evolution does not definitively reconcile the theory of evolution with a belief in God and a higher purpose to our existence. But it does bring these two worldviews

* Biochemists Michael Denton and Craig Marshall have noted: "If forms as complex as the protein folds are intrinsic features of nature, might some of the higher architecture of life also be determined by physical law? The robustness of certain cytoplasmic forms . . . suggests that [they] may also represent uniquely stable and energetically favoured structures. . . . If it does turn out that a substantial amount of higher biological form is natural, then the implications will be radical and far-reaching. It will mean that physical laws must have had a far greater role in the evolution of biological form than is generally assumed. And it will mean . . . that underlying all the diversity of life is a finite set of natural forms that will recur over and over again anywhere in the cosmos where there is carbon-based life."

a little closer together. For me at least, it's much easier and more intellectually satisfying to imagine a God who shaped and guided the evolutionary creative process through higher-order principles than to imagine a God who left the creation entirely up to chance. So hopefully by this point I have convinced you, or at least broadened your horizon to allow for the possibility, that evolution was not a completely random process but that it was guided, constrained, and directed by higher-order principles of nature. This view of life, at least for me, resolves the dilemma of the doctrine of randomness and allows for our continued search for an overall purpose to our existence. And remember: This way of thinking does not fight against the theory of evolution. Even Richard Dawkins, an outspoken atheist and one of Darwin's staunchest modern defenders, acknowledges that nonrandom factors played an important role in evolution.

Now we turn to the issue of humans. If evolution was indeed the means by which a Higher Power created us (through natural principles), then it is certainly possible that we might deduce what this purpose is by carefully considering our evolutionary history and the nature with which it has endowed us.

Chapter Three

Evolution: Our Biological and Psychological Heritage

What is at the heart of human nature? What is it that motivates us? In close relation to this, how does the mind work? Moreover, why does it work in the way that it does? David Hume, one of the great thinkers of the Enlightenment, considered these questions to be of the greatest importance.[1] The answers to them could aid us in restructuring society—perhaps in a way that is unimaginable to any of us—so as to benefit all of humanity. In order to understand the root of human nature, we must retrace our evolutionary footprints.

In a famous address to the American Society of Zoologists in 1964, the biologist Theodosius Dobzhansky proclaimed: "Nothing makes sense in biology except in the light of evolution."[2] Ever since this, biologists have oft repeated some version of it, highlighting that evolution has helped provide a unifying framework for understanding all of biology. It would logically follow that nothing in our study of human nature makes sense except in light of *human* evolution.[3] What are the behaviors and attributes that evolution produced in our ancestors that have been passed on to us? What are the characteristics that allowed, or perhaps compelled, humans to come together to form such complex and intricate social groups? This capacity to join together to form societies is an innate part of us, biologically wired into our nature over tens of thousands of generations. More than any other creature, humans are social beings, and we derive immense satisfaction from our relationships. As I will endeavor to show, our social natures are the root of the more positive aspects of ourselves.

But, as we shall also see, there is a tension in human nature: forming societies means giving up certain rights and requires a suppression of the selfish instincts that exist to some degree in all of us. This capacity to form societies as well as the capacity to be selfish are, in essence, opposing drives. Yet both tendencies have strong biological and evolutionary underpinnings.

Natural Selection

There are a few central principles that, since Darwin, have been foundational to evolutionary theory. First, all biological organisms produce offspring. Second, those offspring vary slightly, both from their parents and from each other. Third, in the natural world, not all offspring survive. Some fall victim to predators; others are not adept enough at scavenging food. Those offspring that survive tend to have inherited variations and traits that help them survive and reproduce. Over generations, traits that are advantageous for survival and reproduction are *selected for* by nature and will continue to be passed on to future progeny. Hence there is a perpetuation of those variations and traits that are most advantageous in the struggle for existence. This phenomenon is known as natural selection. Nature selects *for* those characteristics that assist an organism in surviving; those characteristics that decrease its chances for survival are *selected against* and often eventually eliminated by nature. In Darwin's own words:

> Any variation, however slight and from whatever cause proceeding, if it be in any degree profitable to an individual of any species, in its infinitely complex relations to other organic beings and to external nature, will tend to the preservation of that individual, and will generally be inherited by its offspring. The offspring, also, will thus have a better chance of surviving, for, of the many individuals of any species which are periodically born, but a small number can survive. I have called this principle . . . Natural Selection.[4]

Darwin's thinking was influenced by observations from animals he encountered on his five-year journey aboard the HMS *Beagle.* After leaving the mainland of South America, the ship stopped at the Galápagos Islands. It was here that Darwin observed subtle differences between the giant tortoises of the islands and those of the mainland. He also noted variations in the different bird species that inhabited the islands. Among these were thirteen species of finches that varied in their beaks. His reasoning was that the variations in their beaks allowed the finches to feed in different manners, which could be advantageous, depending upon the availability and abundance of food as well as its location in nooks and crannies between rocks. The simplest way, he concluded, to explain these differences was that each species of bird descended from a common ancestor. Over time, gradual modifications from natural selection led to the subtle differences in species found on the different islands.

Darwin was not the first to put forth a theory that life evolved by gradual means over time. Writing in 1835, the British zoologist Edward Blyth had observed that "among animals which procure their food by means of their agility, strength or delicacy of sense . . . the one best organized must always obtain the greatest quantity; and must, therefore, become physically the strongest and be thus enabled, by routing its opponents, to transmit its superior qualities to a greater number of its offspring."[5] In other words, those animals that are faster, stronger, and smarter have an advantage over their peers with regards to survival.

The phrase "survival of the fittest" was coined in 1864 by the British biologist Herbert Spencer as a way of describing this principle of natural selection.[6] The term "fit" denotes those traits and characteristics that are good for survival. Those traits that confer fitness, or are good for survival and reproduction, depend on the environment.

In the 1970s, the biologist John Endler discovered an interesting example of how different environments can lead to different traits becoming predominant. While studying guppies in Trinidad, Endler and his colleagues observed that the fish were more vibrantly colorful in areas where predators were sparse. Endler thought this might be because predators were alerted to the presence of guppies by bright colors. In this sense, in an environment where predators are plentiful, nature would select against the trait of bright colors by making it more likely for their predators to find them. He tested this idea by relocating a school of darkly colored guppies from an area where predators were abundant to an area where predators were nonexistent. Very quickly, the population evolved to display vibrant coloration, which resulted because females preferred males with brighter colors. Unchecked by predators, the colors of the guppies evolved rapidly over fourteen generations.[7]

Drug-Resistant Pathogens

Another important example of evolution unfolding before our eyes is seen in the microscopic world of bacteria. While most of us alive today take for granted the existence of antibiotic medicines, we only need to go back a few generations to get a sense for what life was like before penicillin. In the 1940s, infections such as pneumonia were often fatal. Charles Fletcher, a young physician of the era who worked on the original clinical trials of penicillin, describes being witness to the remarkable discovery of this first antibiotic:

> It is difficult to convey the excitement of actually witnessing the amazing power of penicillin over infections for which there had previously been no effective treatment. . . . I did glimpse the disappearance of the chambers of horrors which seems to be the best way to describe those old septic wards . . . and could see that we should never again have to fear the streptococcus or the more deadly staphylococcus.[8]

Antibiotics kill bacteria by acting on a specific target. Examples of targets for antibiotics include some structure (like an enzyme or chemical machine) or process (like DNA replication) that is critical for the bacteria to survive and reproduce. Some strains of bacteria, however, have developed clever ways to subvert the antibiotics designed to kill them.

Perhaps one of the most common examples of this is a formidable bug named Staphylococcus aureus. Staph aureus is very common. It is found all over the body of every human being (mostly on the skin and in the nasal cavity and mouth). Generally, these bacteria live in harmony with humans, but certain strains of Staph aureus can cause a broad array of illnesses, ranging from skin infections to life-threatening cases of sepsis. Before penicillin, 80 percent of patients who developed Staph infections died. When penicillin became available, this prognosis improved drastically. Yet, in a cruel twist of evolutionary fate, the use of penicillin led to the selective reproduction of strains of the bacteria that were resistant to penicillin. These versions of Staph aureus produce enzymes that destroy the penicillin compound and evade its antibacterial action. By 1950, over 50 percent of all Staph aureus strains were resistant to penicillin.[9]

In 1959, a more potent antibiotic named methicillin was developed. Methicillin (in contrast to the original penicillin) is immune to the penicillin-destroying enzymes. Yet, at that time, a small portion of the strains of Staph aureus produced a special protein[10] that renders Staph aureus resistant even to methicillin. Over time, as methicillin was used more and more, this evolutionary pressure led to greater reproduction of the type of Staph aureus with resistance to methicillin. These forms of Staph aureus are called methicillin-resistant Staphylococcus aureus, or MRSA. While methicillin is no longer widely manufactured, the term MRSA is now used to describe Staph aureus strains that are resistant to all forms of penicillin. Today MRSA is a big problem, especially in hospitals, where sometimes high proportions of patients as well as healthcare workers have some form of this drug-resistant bacteria living on their bodies. MRSA kills about 700,000 people annually around the world. Until just a short time ago, an antibiotic called vancomycin was very effective against MRSA, but now new forms of bacteria are emerging that are even resistant to vancomycin. Drug resistance among bacteria is a serious problem in infectious disease (and is one of the reasons to avoid everyday hand soap with antibiotics).[11] This back-and-forth struggle between bacteria and more potent forms of antibiotics is a classic example of the principles of evolution being played out before our eyes.

All of us remember more recent examples of real-time evolution from the COVID-19 pandemic. In the United States, the delta and omicron variants of the coronavirus (SARS-CoV-2) displaced all earlier variants during the fall and winter of 2021 because they were so contagious. In evolutionary terms, these variants were so good at surviving and reproducing (even overcoming immunity from vaccines in many cases) that it was only a matter of time before they dominated the coronavirus landscape.

Modern Views of Evolution

Yet, despite overwhelming evidence (including these modern examples of the process unfolding before our eyes), a substantial portion of Americans do not believe in evolution. It seems this is particularly the case with respect to human evolution. For example, based on data from 2009–2014, the Pew Research Foundation reported that only 60 percent of adults believe that humans evolved over time. Some 35 percent of individuals believe that we have existed in our present form since the beginning of life.[12] Granted, it's a bit of a jump to go from observing evidence of bacterial evolution to recognizing that organisms as complex as humans have evolved. But there has now been quite a bit of evidence that similar processes unfolded among *our* ancestors.[13]

Despite growing scientific evidence, evolution—especially with respect to human origins—has been very difficult for many to accept. But why? Why has it been so hard to recognize, as the overwhelming majority of scientists believe, that evolution has shaped human nature?

One issue is the doctrine of randomness. It's hard to reconcile a seemingly random origin of our existence with a universal purpose to our existence. (Hopefully, I resolved this dilemma to your satisfaction in Chapter 2 by presenting a view of evolution that was decidedly *not* random.)

Another issue (the second dilemma) that leads many to resist Darwinism has to do with what evolution—at least as initially understood—implies about human nature. In the struggle for existence, it's easy to understand how selfishness, greed, violence, and the propensity to steal would have been beneficial for survival. Some have used evolution to justify selfish behavior and promote inhumane public policy. Shortly after Darwin published *On the Origin of Species*, several leading sociologists and philosophers (including Herbert Spencer and William Sumner) argued that "survival of the fittest" should apply to human societies as well as nature.[14] Staunch Social Darwinists argued against public programs for the poor because they attempted to subvert the "natural" order. Eugenics, government-forced sterilizations, and genocidal wars have been linked to and sometimes justified by Social Darwinists. In the minds of some, a full acceptance that evolution is responsible for our natures may mean that we surrender to the view that humans are ultimately self-serving creatures and all the terrifying implications that this entails.

Another part of our reluctance to accept evolution has to do with not wanting to give up our freedom. If evolution is responsible for our bodies as well as our behavior, then perhaps we are just organic machines. Perhaps our behaviors are not really chosen by us, but instead are preprogrammed by an exceedingly complex network of genes, proteins, and neurons. This conception of ourselves, in the minds of some, robs us of true freedom. As Edward O. Wilson suggests, "If the brain is a machine of ten billion nerve cells and the mind can somehow be explained as the summed activity of a finite number of chemical and electrical reactions, boundaries limit the human prospect—we are biological and our souls cannot fly free."[15] There is something about this notion that seems to limit our potential and is, to many, off-putting.

Yet despite these misgivings, science continues to mount overwhelming evidence that evolution, at least in some form, is responsible for our natures. I would argue this is not necessarily a bad thing. In a surprising way, when viewed within a certain framework, evolution provides a remarkable, and even optimistic lens through which to view human nature and the potential purpose of life. This framework, at least for me, resolves almost all of the qualms that our society has with recognizing ourselves as the products of evolution.

Genes–The Units of Inheritance (Sort Of)

When Darwin was developing his theory of evolution in the 1800s, very little was known about genetics. While it had long been observed that traits and characteristics were passed from parent to offspring, the way in which this happened wasn't understood. It wasn't until 1953 that Francis Crick and James Watson famously showed how four chemical compounds (called base pairs) fit together to form a twisted ladder structure known as the double helix.[16] This remarkable structure makes the replication and passing on of DNA possible.

DNA—short for deoxyribonucleic acid—is amazing stuff. The entire recipe for building a human body can be spelled out using just these four biochemical letters (abbreviated as A, C, G, and T), repeated in different combinations about three billion times. If you could take the DNA from a single cell in your body and lay it end to end, it would stretch about six feet (or about 100,000 times longer than the actual diameter of an average cell). If you could combine all the DNA from every cell in your body and lay it end to end, it would reach for tens of billions of miles—a simply remarkable notion. This could stretch from Earth to Pluto and back again several times.[17] But because it is so thin, it is packed into a space that is incredibly small. Each time one of your cells divides, the entire DNA of this cell needs to be duplicated. DNA does this with remarkable fidelity; by some estimates, it makes only about one error for every billion letters copied.[18]

The discovery of DNA as the molecular basis for genetics has provided an enormous amount of support to the theory of evolution. Organisms thought to have descended from a common ancestor share remarkable similarity in their genetic material (or DNA).[19]

The central paradox of genetics is that organisms (including humans) can share a remarkable amount of DNA but simultaneously be quite diverse.[20] For instance, all humans share 99.9 percent of their DNA, yet we are all positively unique.[21] Chimpanzees, scientists tell us, are our closest relatives.[22] They share approximately 99 percent of all their genetic material with us.[23] Yet most would argue that humans and chimps are more than 1 percent different. How this is possible is still quite a puzzle. Considering the whole of all our genetic material, oft-quoted figures reveal that we share about 90 percent of our genes with cats,[24] 85 percent with mice,[25] 60 percent with chickens,[26] and a whopping 41 percent with bananas![27] Only about 2 percent of all the genes in our bodies do what we traditionally think of genes doing, that is, give the recipe (or code) for proteins. This means that, according to our current understanding, 98 percent of our DNA does not code for proteins. *Ninety-eight percent!* Why is this 98 percent even there? A small portion of it helps regulate (turn on and off) the genes that do code for proteins. But the vast majority of our DNA doesn't serve any known purpose. For a time, this supposedly nonfunctional portion of our genome was glibly referred to as "junk DNA." Some of it may indeed be junk (genes that have become inactive over the long course of evolution). But an alternative (and in my opinion more likely) explanation is that much of it may simply be DNA whose function we have not yet discovered.

Limits of Genetic Determinism

Genes really are quite remarkable things. But in the last few decades, the power that we have ascribed to them has gotten a little out of hand. This has led to the rise of biological (or genetic) determinism. This is the idea that our biological makeup (as opposed to culture, environment, or personal decisions) is the dominant determinant of our behavior.

There is now a "warrior gene" legal strategy for criminal defenders advocating on behalf of persons accused of violent crimes.[28] (One cartoon depicts an attorney rendering his client's plea before the bench: "Not guilty by reason of genetic determinism, Your Honor.") There is at least one company that will test your genome for a supposed "infidelity gene." You can submit a saliva sample and find out if you (or a potential mate) have the variant of the gene and are therefore likely to cheat in a romantic relationship. All this for the low price of $149! This same company can help you discover whether thrill seeking and risk-taking (the "wanderlust gene") is in your DNA.[29]

But as we have learned more and more about genetics, this notion that genes are the controlling agents of our behavior is less and less convincing. As one scientist (a biologist, no less) writes:

> Numerous commentators expect our genetic endowment to accomplish feats of which it is incapable. . . . Statements

> such as "Understanding the genetic roots of personality will help you find yourself and relate better to others" are, at today's level of knowledge, frankly nonsensical.[30]

One of the critical things we often get wrong when we talk about this subject is this: Genes do not code for traits. They certainly don't code for behavioral traits (such as kindness). But in the overwhelming majority of cases, genes don't even code for physical traits (such as hair color). Some of our misconceptions about this come from grade school lessons of Mendelian genetics. You might remember learning about Gregor Mendel's pea plants and the concepts of dominant and recessive traits. In this elementary lesson, cross-fertilizing two pea plants that have smooth peas (dominant trait) leads to a progeny of plants, where three-quarters have smooth peas (dominant) and one-quarter has wrinkled peas (recessive).

But, as we have come to find out, traits that are inherited in a simple Mendelian fashion are the exception, not the rule. In humans, they are incredibly rare. Human examples of Mendelian inheritance come from disease (sickle cell anemia, cystic fibrosis, and Huntington's disease, to name just a few), not normal physiology. In humans, normal traits that follow a Mendelian inheritance pattern seem to be absent. Even traits that we once thought were inherited in this simple pattern, such as eye color, are now understood to be much more complex.[31]

What genes actually code for are proteins. Proteins are the molecular machines that build and keep our bodies running smoothly. Some proteins fight infection and are called antibodies. Others help provide structure or allow for muscle elasticity (titin). Yet others convey messages through the blood and are known as hormones. How many different types of proteins does a human have? No one really knows for certain. Some experts estimate that our genes make between a few hundred thousand to several billion different types of proteins.[32]

As our understanding of genetics has deepened, the support for genetic determinism has eroded. We have come to recognize that genes don't really act as agents in themselves. They are regulated (turned on or off) by things called transcription factors. In turn, these transcription factors—and here's the death knell for genetic determinism—are regulated by the environment. As the neuroendocrinologist Robert Sapolsky has put it, "It's not meaningful to ask what a gene does, just what it does in a particular environment."[33] What does Sapolsky mean by the "environment"? It can be the environment inside the cell, just outside the cell, or in the world outside the body.

Most scientists assume that genes have *some* influence on our traits—both behavioral and physical—but the way this happens is extraordinarily complex[34] and very indirect. There are very few examples in humans where a known genetic variation has quantitatively been linked to a behavioral response. Even when we do find genetic variation that can be quantitatively linked to a behavior, these estimates are very imprecise. Genes do indeed shape and influence our behavior; but they don't determine it. In other words, there is no gene for selfishness or aggression; likewise, there is no gene for kindness or cooperation. Clearly genes play some role in our behavior. But the reality of human psychology is increasingly seen to be worlds away from a sort of genetic behavioral determinism. This is an important principle, and one that is critical to later discussions.

The Levels of Selection

Let's return to the principle of natural selection. A critical concept is what some biologists refer to as the question of the *level of selection*. This question asks: On what level of the biological hierarchy does natural selection operate? Genes? Cells? Individual organisms? Groups? In the natural world, where not all biological entities can survive, does nature select the fittest gene? Or does it select the individual organism with traits that make it best adapted to survive? What about the fittest family? Given that family members share genes with each other, is it possible that families compete against other families, with the most cohesive family producing more progeny? As we shall see, this is a critical question, and its answer can have profound implications with respect to the behavioral traits that evolution endows to a given species.

Individual Selection

When Darwin first proposed the theory of evolution, it was thought that the individual organism was the principal level or unit of selection. In other words, nature selected favorable traits and features in a species by virtue of the fact that *individuals* with such traits were more likely to survive and propagate these traits to the next generation. Hence, the default level of selection since the time of Darwin has mostly been the *individual level.* Elliott Sober and David Sloan Wilson give the example of the ability to run fast within a population of zebras:

> Zebras, for example, will gradually increase in running speed because faster zebras do a better job evading predators. Faster zebras are *fitter*—they are more able to survive and tend to have more offspring than slower ones. If offspring resemble their parents, the frequency of fast zebras—the proportion of good runners in the herd—will increase. Notice that a zebra that runs fast benefits itself—not other zebras, not lions, not the whole ecosystem.[35]

Sober and Wilson continue to explain how individual selection will produce a certain kind of behavioral trait—selfishness—and reduce an opposing trait—altruism.

> In this example, natural selection favors those who help themselves. It therefore appears that helping other individuals to survive and reproduce at the expense of one's own survival and reproduction is the very thing that natural selection will eliminate. In short, natural selection [on the individual level] appears to be a process that promotes selfishness and stamps out altruism.[36]

Kin Selection

It is now established doctrine that natural selection operates on the level of the individual. However, it turns out that this is not the complete picture. As Darwin himself wrote in *On the Origin of Species*, he noticed a potential flaw, one that "at first appeared insuperable, and actually fatal to [the] whole theory [of evolution]."[37] How is it, that in some species, particularly in those of intricate social order, traits and behaviors would arise that are actually harmful to the individual but beneficial to the community?

For instance, at the threat of attack, honeybees will sting an intruding animal or human. The stinging apparatus of the worker honeybee is barbed: it has rows of lancets, pointing outward, like saw-toothed blades (see Figure 3.1). This means that if the bee stings an animal with thick skin (like a human), the stinger acts as an anchored screw. Once the stinger is inserted, it cannot be withdrawn without ripping out vital internal organs that are attached to it. The outcome is death for the stinging honeybee. Yet why would this evolve? Perhaps the ability to sting an intruder, despite resulting in certain death for the individual, was advantageous for the family of bees. This would seem to be the case, as honeybees rarely sting when they are not near the hive. Additionally, if in defense of the hive, a honeybee will not only sting a perceived intruder, but will also emit an attack hormone, or pheromone. This hormone signals danger to other bees and alerts them to come and help defend the hive.[38] Hence, natural selection may have favored the development of the stinging mechanism (as well as the emission of pheromones) to benefit the community of honeybees, despite its fatal outcome for the individual.

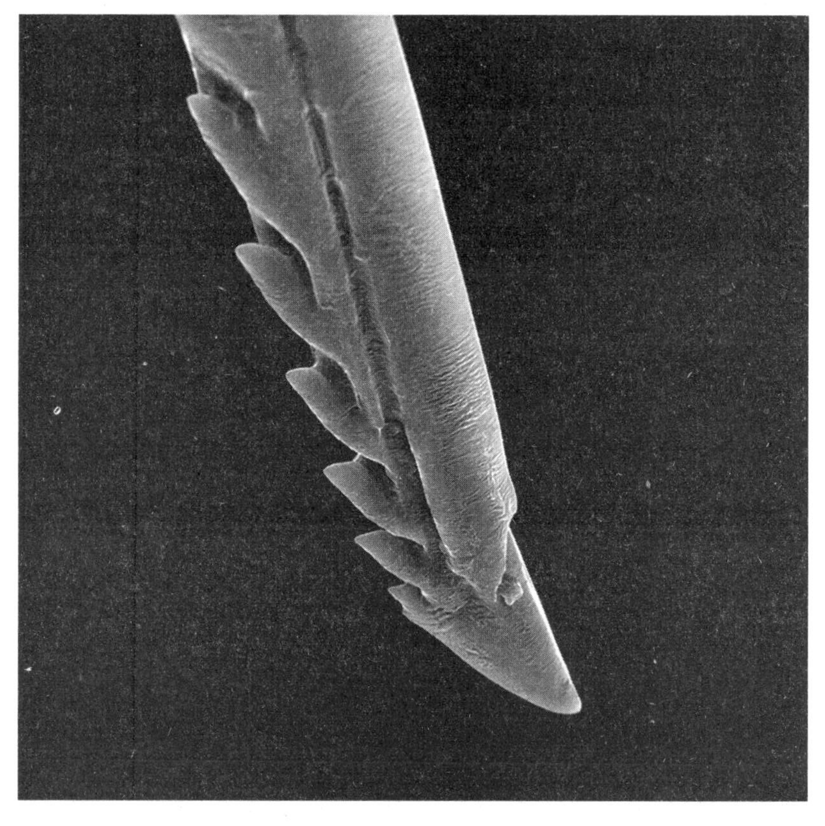

Figure 3.1. The barbed stinging apparatus of a honeybee.

The example that initially puzzled Darwin was not the stinging mechanism of honeybees, but the fact that in hives of honeybees the female worker bees are completely sterile: they cannot produce their own offspring.* Each hive of honeybees has approximately 60,000 infertile female worker bees, with a few hundred male drones, and a single female queen. If the ability to reproduce is such a critical concept in evolution, why would hives have evolved to produce tens of thousands of infertile female workers? These worker bees toil their entire lives to support a few hundred males and a single female who are afforded the opportunity to reproduce—a dramatic display of biological altruism! Why would they exhibit such behavior? The answer, sketched out initially by Darwin himself, was that natural selection sometimes operates at levels other than the individual level. "In social animals," he wrote, "[natural selection] will adapt the structure of each individual for the benefit of the community."[39]

In honeybees, these sterile female workers are all close relatives of the queen and male drones. They share much of the same genetic material. Hence, their traits are passed on through the queen. Their immortal genes live on in the bodies of their thousands of nieces and hundreds of nephews. Furthermore, honeybees are tasked with building a huge hive with sufficient honey to last through the winter and provide enough nourishment and energy for the larvae of subsequent generations. This task is immense and requires a tremendous amount of cooperation. A hive of bees that does not cooperate very well would surely not be as fit as a hive of bees that works harmoniously together. Hence, we can explain the related phenomena of sterile female worker bees and the high degree of cooperation within the hive by recognizing

* This is similar to other caste insects, such as ants, wasps, and termites.

that natural selection might operate on a family level in addition to an individual level. This is called *kin selection*.

The Harvard biologist Edward O. Wilson, one of the world's leading experts on social insects, describes this phenomenon:

> Imagine a network of individuals linked by kinship within a population. These blood relatives cooperate or bestow altruistic favors on one another in a way that increases the average genetic fitness of the members of the network as a whole, even when this behavior reduces the individual fitness of certain members of the group. The members may live together or be scattered throughout the population. The essential condition is that they jointly behave in a way that benefits the group as a whole, while remaining in relatively close contact with the remainder of the population. This . . . is *kin selection*.[40]

Part of what makes kin selection possible is that individuals who are closely related share genetic material. These genes, in turn, work in remarkably complex ways to influence the traits and characteristics of the next generation. A set of genes that influences the individual to behave in ways that benefit the family will likely be passed on, even when it might be at some cost to the individual.

This concept of kin selection was initially worked out in the 1930s by two biologists—R. A. Fisher and J.B.S. Haldane. One of them famously joked that he would willingly give his life for either two of his brothers or eight of his cousins. But there is truth in jest, as formalized in 1964 as Hamilton's rule: the more closely related two biological organisms are, the more genetic material they share, and thus the greater the chance that one will be willing to sacrifice for the other. This is one of the key ways that evolution developed the capacity for altruism within us. Again, the language of Edward O. Wilson is helpful:

> Consider, for example, a simplified network consisting solely of an individual and his brother. . . . If the individual is altruistic he will perform some sacrifice for the benefit of his brother. He may surrender needed food or shelter, or defer in the choice of a mate, or place himself between his brother and danger. The important result, from a purely evolutionary point of view, is loss of genetic fitness—a reduced mean life span, or fewer offspring, or both—which leads to less representation of the altruist's personal genes in the next generation. But at least half of the brother's genes are identical to those of the altruist by virtue of [sharing the same parents]. Suppose, in the extreme case, that the altruist leaves no offspring. If his altruistic act more than doubles the brother's personal representation in the next generation, it will . . . increase the one-half of the genes identical to those in the altruist, and the altruist will actually have gained representation in the next generation. Many of the genes shared by such brothers will be the ones that encode the tendency toward altruistic behavior.[41]

Kin selection and Hamilton's rule are now established doctrine in biology. We can even see evidence of these concepts in humans. In one social science experiment, human participants were asked how willing they would be to help others in different hypothetical situations, ranging from donating a kidney to giving someone money to throwing oneself on a grenade. The more closely related the study participants were to the potential beneficiaries, the more willing they were to help.[42] In another study, people would spend longer in an uncomfortable position if it helped family members compared to strangers.[43] In another study, the amount of money that migrant workers sent back to their country of origin was linked with the genetic relatedness of the recipients.[44] Maxwell Burton-Chellew and Robin Dunbar reported a strong connection between how closely related you are to someone and the list of people you would ask for help if you were the unfortunate victim of a sudden catastrophe.[45] Overall, Hamilton's rule applies to humans: we are more likely to behave altruistically toward those with whom we share genetic material.[46]

Group Selection

As its name implies, group selection is the operation of natural selection on a group level.* Group selection was not always seriously considered as a force in evolution (and is still controversial). In fact, in the 1960s it was derided and dismissed with an almost religious zeal. Part of this was because of erroneous and flawed observations that some scientists used to support group selection; another part was that the original definition of group selection was overly broad. In recent years, however, thanks to the patience and meticulous work of David Sloan Wilson (and other scientists), a nuanced form of group selection is resurfacing and gaining some acceptance. More scientists are recognizing that there are specific circumstances in nature when group selection is a plausible explanation for the emergence of physical or behavioral traits.

* As often happens in science, much of the controversy surrounding group selection involves definitions. The definition of a group in group selection theory has been part of the problem. On a basic level, a group can be just two or more individuals. In this case, reciprocal altruism (or direct reciprocity, discussed in Chapter 4) is a special case of group selection, where a pair of cooperative individuals outcompetes other pairs of less cooperative individuals.

One classic example comes from the work of William Muir and his colleagues at Purdue, who were trying to breed chickens to become more productive in laying eggs. In order to do this, Muir began an experiment in evolution. In nature, chickens live in flocks. In the poultry-producing industry, they are housed about six to nine hens per cage. The experimenters would select the two or three most productive chickens from each cage and then put them in a group together. The thinking was that fertile chickens would produce fertile chickens, and that by grouping these chickens together in subsequent generations, the trait of egg productivity could be enhanced. But the experiment backfired. Subsequent generations of cages of "super chickens" started laying fewer eggs. After about five generations, a cage starting with nine hens ended with only three. Six had been murdered, pecked to death by the survivors. What's more, the survivors had plucked each other's feathers off!

It turns out that the most productive hens in each cage achieved their productivity by being bullies to the other hens and hoarding resources. So, unwittingly, the experimenters weren't only selecting for the heritable trait of egg-laying productivity; they were also selecting for the heritable trait of bullying. As the hens were bred to be more and more aggressive, they spent so much energy and effort struggling just to survive that they couldn't expend much energy producing eggs.

The correct approach, as revealed in a parallel experiment, could have been predicted by applying the principle of group selection. Instead of breeding for *individual* hens that were successful, the experimenters now bred for *groups* of hens that were successful. The offspring from the most successful cages were then grouped for the subsequent generation. As a result of this new enlightened approach, after a few generations, egg-laying productivity increased 160 percent.[47] Muir's "super chickens" are a great example of group selection at work (albeit with artificial, rather than natural selection). The experimenters selectively bred the *group* of hens that was most productive, which gave rise to even more productivity in future generations. A similar observation has been made among spiders.[48]

Darwin himself had noted that traits that might be disadvantageous at the individual level (like altruism) might be very helpful at the group level if possessed by a majority of the group's members. Writing in 1871, he observed:

> It must not be forgotten that although a high standard of morality gives but a slight or no advantage to each individual man . . . yet . . . an increase in the number of well-endowed men and an advancement in the standard of morality will certainly give an immense advantage to one tribe over another. A tribe including many members who . . . [possess] a high degree of . . . patriotism, fidelity, obedience, courage, and sympathy, were always ready to aid one another, and to sacrifice themselves for the common good, would be victorious over most other tribes, and this would be natural selection.[49]

So it is with many traits, especially those that are behavioral in nature. A trait that may seem like a weakness (like altruism) when competing *within* the group may be a great strength when it is possessed by a majority of the group's members. Cages of cohesive hens will outcompete ones comprised of bullies that peck each other to death. David Sloan Wilson, writing with Edward O. Wilson (no relation), summed it up nicely: "Selfishness beats altruism within groups. Altruistic groups beat selfish groups. Everything else is commentary."[50]

A growing amount of evidence suggests that group selection did indeed have a role,[51] especially in the evolution of human behavior.* The group-selection theory can elegantly explain several of our behaviors that are not commonly found in the animal kingdom: our ability to cooperate extensively, our propensity to form friendships with unrelated individuals, and our capacity to be kind to people even if we are unlikely to interact with them in the future (sometimes anonymously). Mathematical simulations by the economist Samuel Bowles have even suggested that group selection could have been a major force that shaped human behavior prehistorically.[52]

It is beyond my purpose here to resolve the academic debate on the extent to which group selection may or may not have shaped human nature (it's still fairly controversial).[53] What is generally accepted, however, is that kin selection *did* occur and has invariably shaped human nature in critical ways. In our evolutionary history, natural selection has operated on more than one level.

Multilevel Selection

In theory, as some biologists have noted,[54] natural selection can operate on any level of biological entity (genes, cells, individuals, family groups, etc.) where there is variation of heritable traits in a population, and where these traits enhance the entity's ability to survive and reproduce. This realization has led to the development of *multilevel selection theory*, which states that, depending on the context, natural selection can act on various levels, including the gene, the individual, the kin group, or even larger groups. Multilevel selection theory was developed in part to quell the intense (and sometimes acrimonious) debate among biologists about what the most meaningful level of selection is. On the one

* The idea of group selection becomes more acceptable to its critics when it is recognized to have operated on a *cultural* level rather than on a *genetic* level.

hand, some biologists have argued for a gene-centric view.* The single gene is the ultimate level on which natural selection operates. On the other hand, other biologists have argued that other levels—including groups in some circumstances—are at play.**

As we shall discuss in later chapters, these different levels of selection have created competing natures within us. The most obvious examples are selfishness and altruism. Selfishness is clearly the result of natural selection operating on the individual level; yet altruism—at least at the deepest level—arises from kin selection, as parents rear offspring who represent the perpetuation of their own genetic material. In like manner, siblings act altruistically toward each other as well as extended family. The capacity for friendship, even between individuals with no close biological connection, is a powerful force that arises from natural selection operating on a level higher than the individual.

The capacity for aggression and conflict is also a result of natural selection at the level of the individual. Yet again, the strongest forms of cooperation come through natural selection operating on a higher level. This has allowed for groups of humans to bond together and accomplish shared goals and tasks that would be impossible for any single person to undertake alone. The capacity for greed and generosity can be treated in the same way, each arising from natural selection operating on different levels. In like manner, we could consider the capacity for loyalty and betrayal. These different levels of selection have created within us opposing capacities across a range of social behaviors and abilities. This was noted by the evolutionary biologist W. D. Hamilton, a pioneer in this field:

---

* For instance, the biologist Richard Dawkins has argued quite loudly for a gene-centric view.

** David Sloan Wilson and Edward O. Wilson have argued that other levels are at play in certain circumstances.

> There are many traits, like resistance to disease, good eyesight, and dexterity, which are clearly beneficial to individual, group, and species. But with most traits that can be called social in a general sense there is some question. For example, as language becomes more sophisticated there is also more opportunity to pervert its use for selfish ends: fluency is an aid to persuasive lying as well as to conveying complex truths that are socially useful. Consider also the selective value of having a conscience. The more consciences are lacking in a group as a whole, the more energy the group will need to divert to enforcing otherwise tacit rules or else face dissolution. Thus considering one step (individual vs. group) in a hierarchical population structure, having a conscience is an "altruistic" character.[55]

Kin selection is not the only explanation or mechanism for altruism or other prosocial behaviors. In future chapters, we explore other variations of evolutionary mechanisms for prosocial behavior. But an important argument, and one to which we will return later, is that the strongest forms of prosocial behaviors have come about through kin selection. In other words, we are most biologically preprogrammed to cooperate with, act altruistically toward, and sacrifice for those with whom we most share genetic material: our families. This comes about by virtue of kin selection.

In the following three chapters, we explore in detail a few examples of how natural selection and other principles of evolution have endowed humans with seemingly opposite tendencies. These opposing characteristics are critical to laying a framework that provides a higher-order purpose for our existence.

Chapter Four
Selfishness and Altruism

That which is not good for the beehive cannot be good for the bees.[1]

In 1985, Kenneth Lay founded the energy company Enron, headquartered in Houston. Five years later, Lay hired an impressive young consultant named Jeffrey Skilling to work as a leader in Enron's financial division. By all measures, Skilling was talented: he had graduated in the top 5 percent of his class at Harvard Business School, where, according to his peers, he had displayed pure genius and a knack for imposing refreshingly new ideas on top of existing frameworks. After graduating, Skilling had gone to work for a prestigious consulting company, which at the time was doing work with Enron. In the course of his work, Skilling attracted the attention of Lay, who was so impressed with his acumen that he created a new division for him to lead.

Skilling swiftly ascended the corporate ranks and in 1996 became Enron's chief operating officer, which made him second only to Lay in the company's leadership structure. By that point, Skilling had developed a reputation at Enron as arrogant and intimidating. Employees later recounted an episode when, upon leaving the parking garage in his car at the end of the workday, Skilling impatiently raced around a line of cars (presumably all driven by his own employees) while giving them the middle finger.[2]

During his rapid climb up Enron's leadership ladder, Skilling had laid out an aggressive management strategy and promoted an almost ruthless corporate culture. Skilling's view of the business world, and the world at large, was that life is a struggle of the fittest individual.* People, in Skilling's view, are primarily motivated by fear and individual gratification, and the way he shaped the corporate culture at Enron reflected this philosophy. For example, he established the Performance Review Committee, or the PRC—a mandatory evaluation process that occurred twice per year. Colloquially called the "rank and yank," it consisted of a "360-degree" evaluation, meaning that people "beneath, above, and horizontal" to you on the management tree would evaluate your performance.[3] The employee up for review was required to obtain evaluations from five people. In addition, however, people who knew (and perhaps disliked) the employee could submit unsolicited feedback. Following the solicitation of feedback, supervisors would gather in a PRC meeting (without the person being evaluated) to hash out the process. These PRC meetings, recalled former employee Brian Cruver, could get "downright nasty," adding, "People were either fighting for you or against you as [your picture] smiled on the slide-show screen. There was no statute of limitations, either. You could dig up any old dirt you wanted and fling it across the table at the target employee."[4]

Eventually, the employee up for review would be ranked from one to five along a forced curve. This meant that 15 percent of the people had to receive a rating of five (the worst category) and 15 percent had to receive a rating of one (the best category). Employees in the bottom

* Coincidentally, Skilling's favorite book was *The Selfish Gene*.

15 percent were "redeployed," meaning they were sent to find another job within the company, but with a scarlet letter on their file that let everybody know they had been appraised in the bottom 15 percent during a PRC review. What really happened, according to one former employee, was this: "You get fired, only in a really slow and painful way."[5] That was Skilling's corporate version of "survival of the fittest."

Initially, Enron's strategies, ruthless PRC and all, seemed to work beautifully. Throughout the 1990s, Enron was lauded as one of the most innovative and successful companies in the United States. Headlines praised Lay and Skilling's pioneering approaches to the business. Skilling was eventually made CEO in February of 2001.

Yet, as Enron climbed the charts of Wall Street, the pressure of the PRC process led employees and the company as a whole to take shortcuts to get ahead—to be less than honest. In an effort to survive, employees inflated the value of financial transactions and exchanges to make it seem as though their contributions to the company were worth more. This was done in hopes of garnering a spot in the coveted top 15 percent in PRC reviews (or, to avoid being branded in the bottom 15 percent and seeing their career at Enron come to a slow, painful death). Over time, as the individual employees of the company continued to inflate the value of the financial transactions, the perceived value (and stock price) of Enron soared. Meanwhile, the company's actual assets were much lower than believed. Between 1997 and 2001, Enron had inflated its income earnings reports by approximately $586 million. In August 2001, just six months following his promotion, Skilling abruptly resigned as CEO. Enron filed for bankruptcy a few months later, collapsing with breathtaking speed. Thousands of employees lost their jobs and retirement savings. Investors lost billions. After a protracted trial, Skilling was convicted of twelve counts of securities fraud, five counts of making false statements to auditors, one count of insider trading, and one count of conspiracy. He was sentenced to twenty-four years in prison and ended up serving twelve years before being released to a halfway house.

What can we learn about human nature from the Enron fiasco? In complex social systems, it seems, groups that pit individuals against one another in a purely Darwinian way seem to fail. Enron maximized selfishness, greed, and fear in its employees, which encouraged them to behave in ways that served them well in the short-term but were ultimately catastrophic for the well-being of the company in the long-term.

In attempting to establish a system that mimicked Darwinian natural selection, Skilling failed to realize that natural selection operates on a higher level than just that of the individual. Groups comprised of unselfish individuals will outcompete groups comprised of selfish individuals, and the power of a cohesive and cooperative group is always superior to the power of an individual.

The case of Enron poignantly exemplifies how competition operating on a purely individual level, and the selfishness it induces, can paradoxically lead to worse evolutionary fitness. To be sure, the prison time served by Skilling and others may not be what most biologists have in mind when they think of reduced evolutionary fitness. However, it is not beyond reason to conclude that in prehistoric societies—such as those in which our ancestors evolved—serious violations of social and ethical codes of conduct, analogous to the crimes committed by Enron's executives, might have led to ostracism or even capital punishment.[6] As human social groups grew increasingly complex, societies developed ways to keep cheaters in check. The development of punishment, guilt, shame, and remorse were likely all critical to keeping would-be cheaters from dominating society.[7]

As the Enron story suggests, natural selection at the individual level does not always lead to improved evolutionary fitness in highly complex social networks. It also doesn't explain the behaviors of organisms that form intricate societies. Darwin himself noted this as he observed the complex networks created by social insects such as ants and bees. Why would evolution have created creatures such as worker bees that have no capacity to reproduce and yet spend all their lives helping others?

Related to this is the problem of altruism. In a strictly biological sense, altruism refers to behavior that is costly to the individual performing it, but beneficial to others. How does natural selection on an individual level produce altruism? How would it be that one organism could self-sacrifice for the benefit of another? What evolutionary purpose would this serve? Yet examples of acts of altruism abound, both in humans and animals.

Birds, especially in their role as parents, have been observed to risk their lives for their offspring. When a potential predator approaches the nest, the parent bird will often feign injury or fly low to the ground, sometimes even landing right in front of the predator. This behavior distracts the predator's attention away from the helpless offspring, but at substantial risk to the parent.[8]

Vervet monkeys behave similarly, sounding alarm calls for the good of the group at the approach of a predator. This action increases the chance that the group will survive but also that the sentinel monkey will be attacked.[9]

This quandary, the problem of altruism, is solved in part by the levels of selection. Natural selection can operate at more than the individual level. Especially when organisms share a high proportion of genetic material, altruism can be performed by one individual for its relative to enhance the survival of such shared genetic material. This is another way to describe kin selection: individuals are biologically primed to be altruistic to their family members.

Two Shipwrecks at the Auckland Islands

In humans, however, close kinship is not required for altruistic behavior. Friendship among nonkin is, at least in modern society, extremely common. At some point in our evolutionary past, it became increasingly important for humans to behave in altruistic ways toward each other and to work together in cooperative groups. While the most basic social group is the family, it's also very likely that it was increasingly important for nonkin to work together. The Yale sociologist Nicholas Christakis conducted a fascinating analysis of two shipwrecks at the same time and place that show the inherent principles of society. However, these stories also show the contrasting survival value of selfish and altruistic groups.[10] Drawing upon Christakis's analysis of these stories, we can easily imagine how altruism came to be favored in our evolutionary past.

In 1864, two Australian vessels, the *Grafton* and the *Invercauld*, shipwrecked on opposite sides of the Auckland Islands, an archipelago 300 miles south of mainland New Zealand. The *Grafton* carried five passengers and wrecked on the south end of the islands. After twenty months, all five of the crew of the *Grafton* survived this horrifying ordeal. Two of them kept journals (using seal's blood to write after their ink ran out) that were later published. The *Invercauld*, for its part, carried twenty-five passengers and wrecked on the northern end of the island. Nineteen of these twenty-five made it to shore after the wreck, but only three of them survived. Although both groups overlapped for a year as shipwreck victims on the same archipelago, they each knew nothing of the others' presence.

These two shipwrecks allow for an examination of human nature from the lens we have been crafting with respect to our opposing tendencies. Notably, they can help us imagine how similar circumstances and experiences in our prehistoric past may have favored the development of tendencies toward altruism, cooperation, and cohesion that opposed those of selfishness.

The *Grafton* was intended as a business venture to explore the potential of mining and sealing on the Auckland Islands. The officers were François Raynal—a businessman who had helped fund the expedition—and Thomas Musgrave, the captain. As the ship neared the islands, around New Year's, a heavy gale forced them to wait out the storm and drop anchor. However, the storm was so intense that it tore the chain connecting the ship to its anchor, and the *Grafton* struck a rocky beach.

As Christakis notes,[11] the tenor and tone of the whole ordeal was established as the five men made their way to shore. Raynal had been injured in the shipwreck and required assistance to make it to land. In a supreme act of courage, Musgrave carried Raynal on his back as he swam to shore in chilly waters. As Musgrave approached the shore, he cried out to another crew member that he could no longer carry Raynal. At this point, the crew member jumped in to carry Raynal the rest of the way. This initial act of altruism helped unite the group into a cohesive team, preparing them to weather remarkable ordeals for the next twenty months.

They were fortunate enough to be able to start a fire on that first day on shore. Raynal—who was too injured to move about much—tended the fire while others went off in search of anything that might be useful. Unfortunately, they returned empty-handed. As they gathered around the fire and the reality of their fate sunk in, Captain Musgrave began to weep loudly.[12] The captain later wrote of his thoughts upon realizing the gravity of their situation: "God only knows whether we are all to leave our bones here . . . And what is to become of my poor unprovided-for family? It drives me mad to think of it. I can write no more."[13]

After scavenging the wrecked ship and obtaining some timber and other supplies, they were able to pitch a tent further inland. Within a few days, they managed to kill a seal, which they used for food. The weather, as is typical for the Auckland Islands, was overcast and rainy for their entire first week. Then, on Sunday, January 10, the sun finally emerged. The men saw in this an omen of good fortune. Raynal recorded: "In this moment of peace and benediction, after the terrible trials we had undergone, we all . . . felt in the bottom of our hearts the awakening of an irresistible need of devotion."[14] This sentiment was reinforced when Musgrave found a Bible among the supplies that had been salvaged from the ship. Raynal recorded: "We begged him to read us some fine passage from the Gospels . . . and ranged in a circle round him before the tent, we listened with the deepest attention." When he read of Christ's teaching to His disciples to love one another, all the men "burst into tears."[15] A sentiment of unselfishness and brotherly love had marked the beginning of their shipwreck, which played a large role in their cohesion and the eventual survival of the whole group.

About a month following shipwreck, it was clear that the dynamic of leader-crew relationship was changing. Though Musgrave was captain while on board, he technically no longer had authority over the crew on land. There were rumblings of disobedience. Up until this point, the men had worked together, as Raynal recorded, "in peace and harmony—I may even say in true and honest brotherhood."[16] Both Musgrave and Raynal realized that if they were to survive, they would need to continue to work together.

To quell rising contention, the men decided to form a sort of democracy and elect a chief of the family. All the men agreed to this, with the sailors requesting a provision that this man could be recalled if he ruled poorly. The men were unanimous in their support of this system and a sort of legal contract was recorded on an empty page in Musgrave's Bible. The men clasped hands and solemnly made an oath to obey the contract and, upon finishing this ritual, they held a vote. Raynal nominated Musgrave. His election was unanimous, and Raynal unselfishly volunteered to do the cooking for the first week of this new social order, as a means of soothing any unkind feelings that were still raw.[17]

After about a year on the island, the men realized that waiting to be rescued was all but hopeless. At this point Raynal convinced them to create a plan to rescue themselves. The men fashioned a bellows from which they forged nails, pegs, and bolts. They stitched sails that they would attach to the small dinghy that had survived the wreck. They worked together to produce new timber, a very difficult task due to the twisted, knotted trees that were native to the island.

Shortly after finishing the boat and testing its abilities in the harbor, Musgrave realized that five men were too many to travel such a distance (300 miles) in such a small boat. After much deliberation, it was decided that two sailors would stay on the island while Musgrave, Raynal, and another would venture forth upon the rough water and hope to reach New Zealand. Musgrave promised that he would return and rescue them as soon as he found the means to do so. Their parting was heartfelt and sorrowful. As they bid farewell, "they joined in prayer to God, imploring his assistance for those who, in a frail bark, were about to confront a stormy sea, and those who remained on the rocky isle, to wrestle alone against want and despondency."[18] As one historian noted, these "five men had been comrades for the past twenty months. . . . They had shared the same sufferings and struggles; they had worked in close brotherhood for the good of them all. Because of conscientious leadership, resourceful technology, unstinting hard work, and an outstanding spirit of camaraderie, they had survived unimaginable privations."[19]

After five grueling days and nights at sea, the tiny dinghy finally arrived at Stewart Island of New Zealand. As soon as he was able to make arrangements, Musgrave and others returned to the islands and rescued the other two crew members. As they left the Auckland Islands, however, Musgrave spotted a column of smoke. Unable to stand the thought of leaving another unfortunate soul to die on this island, Musgrave and his crew spent considerable time and effort searching for others stranded in the vicinity. On the north end of the archipelago, they found somebody who had recently died, and later identified him as a former sailor of the *Invercauld*. They took the effort to bury the man, say some prayers, and light a fire so that others nearby might detect their presence. But a full search of the island was impossible given the weather, their small supplies, and the dense brush that made further voyage into the heart of the island impossible. For Musgrave, this was most unsatisfactory. As he later wrote, "I confess the doubt [of having left someone] torments me. The thought that some poor wretch should be left upon it to suffer what *we* suffered pursues me incessantly."[20] After reuniting with his family, Musgrave returned a third time to the Islands on a steamer, in part to satisfy this haunting torment that there may be some abandoned castaway holding out hope of rescue.

A number of books have been written about the remarkable experience of the survivors of the *Grafton*. There are several factors that historians have noted were important in the remarkable fact that all five men survived: the leadership of Musgrave and Raynal, the ingenuity of the crew, the fact that they were able to save some supplies from their wrecked ship. One factor is clear, however, and that was their cooperation and altruism. Certainly, they had their disagreements, sometimes heated ones. But their collective altruism, as evidenced by that first unselfish act of rescuing an injured comrade, was critical in contributing to the collective survival of them all.

Unfortunately, this was not the case with the *Invercauld*. This ship, captained by George Dalgarno, with first mate Andrew Smith, was headed from Australia to Peru when it wrecked on the north end of the Auckland Islands in May 1864, just five months after the wreck of the *Grafton*. The crew of the *Invercauld* found themselves at the foot of cliffs in the middle of the night, with each man being forced to make his own way through the water toward the shore while at the mercy of the waves and surf. Six men died that first night, leaving alive nineteen others who anxiously awaited daylight. They were able to save two pounds of pork and a similar amount of bread from the wreck, but not much else. Luckily, one of the men had some matches in his pocket, and (despite the fact that the matches were waterlogged) they were able to build a fire within a day of shipwrecking.

They stayed at the foot of the cliffs for a few days before a small group of them ventured to climb the cliffs and found some wild pigs, one of which they were able to kill and eat. Upon learning that prospects of survival might be better atop the cliffs, all the men mounted the cliffs, except for one who was injured in the wreck. This man was left behind and died within a few days. This first act of selfishness—leaving a wounded man to die—epitomized the dynamics of the *Invercauld* survivors.

A few days later, a group of four went in search of another pig. After pursuing the pig for most of the day, only three men returned to camp. The fourth, who had become exhausted, was left behind and found dead the next day.[21]

In about a month's time, the survivors stumbled upon two uninhabited houses and found a few tools, including a hatchet. They were subsequently able to fashion a boat, which allowed them to search out and travel to other nearby islands.

At this point in time, the social dynamic was characterized by frequent splintering and occasional reuniting of smaller groups. There was a lack of leadership and group cohesion. As a result, many men died while they were off alone or in pairs. After some months on the island, there were only three still alive: Captain Dalgarno, First Mate Smith, and Robert Holding, who proved to be a very resourceful seaman. Eventually, they made a better boat by tearing down one of the houses and using the hardwood planks for materials. Unfortunately, one night a storm blew in over the island, and when the three men arose the next morning, they were crestfallen to find that their boat had been swept away by the wind and the surf.[22]

In time, like the survivors of the *Grafton*, the remaining three crew members of the *Invercauld* began to form some sense of group cohesion (though the relations were often strained). The three survivors made a shelter-house, approximately ten feet square. They used seal skins for blankets. As Smith later recalled, "When finished, we found it very comfortable, I can assure you. The bottoms of our beds were of seal's skins stretched upon a stretcher, and then we covered them with some withered grass, and for blankets we used seal-skins."[23] The sealskins were also used in the repair of their clothing and shoes, for as Smith attested: "Our clothes were by this time all worn out, and we had to repair our shoes every week, or make new ones."[24]

As this group of three struggled to survive, the dynamics of leadership had clearly shifted. Holding seemed to be the most capable of survival. Eventually, much of the command and leadership of the group was ceded to him, despite the fact that he was a seaman and the other two were officers. Using their small boat, the group traveled from island to island within the archipelago, always searching for better sources of food. Finally, as they were hunting in the north end of the Auckland Islands, they spotted a Spanish ship, the *Julian*. Despite the fact that a contagious illness was infecting many of the sailors aboard, the men were eager to take their chances on board the ship.

The events surrounding their rescue exemplify the spirit of disunity that had characterized their twelve-month ordeal. As the *Julian* approached, Dalgarno turned to Holding and snapped: "Don't *you* speak to them. *I* will be the one who speaks."[25] Apparently Dalgarno was anxious to reclaim his authority as the presiding officer over a seaman. Once again aboard a ship, where the distinctions of rank were more enforceable than they had been on the island, Dalgarno seemed satisfied that he and Smith received superior treatment while Holding was relegated to dwell with the common seamen. The three surviving *Invercauld* passengers, along with the crew of the *Julian*, landed at Callao, Peru, on June 26, 1865. Initially hesitant to publish anything about their experience, Dalgarno and Smith eventually each wrote a description of the ordeal they had suffered. But their relationship had deteriorated markedly by this point.[26]

In terms of our understanding of human nature and the evolutionary forces that shaped it, the twin wrecks of the *Grafton* and *Invercauld* are about as close as we can come to a natural experiment. They offer a valuable case study with a written record of which behaviors and tendencies in human nature might have been beneficial for survival. Some historians have argued that the survival of the crew of the *Grafton* was due to the supplies and material they were able to salvage from their wrecked ship. The *Invercauld*, they argue, was completely obliterated within twenty minutes of shipwrecking[27] and the crew salvaged only two pounds of pork and a similar amount of bread. However, the *Invercauld* survivors soon stumbled upon already-built shelters with tools, timber, and other materials. Indeed, it seems that the ethos of each shipwrecked group was crucial as a factor in their survival success or failure. The men of the *Grafton* worked together and were dedicated to the overall survival of the group, as exemplified in the first heroic act of saving the injured Raynal. In contrast, there emerged among the crew of the *Invercauld* an every-man-for-himself approach,[28] as demonstrated in leaving the wounded man to die on the first day of shipwrecking.[29] The cases of the *Grafton* and *Invercauld* exemplify the ways in which natural selection might have favored the group rather than the individual under different contexts in our evolutionary past. While the members of the *Grafton* were not related, it is easy to imagine how the principles exhibited by this crew—principles of comaraderie, cooperation, and unity—would especially favor the survival of kin groups.

◆

Among the crew of the *Grafton*, the initial extraordinary act of altruism, where Musgrave carried the injured Raynal on his back as he swam to shore, helped define this spirit of cooperation that allowed for their collective and miraculous survival.[30] Yet many other examples, no less heroic, can be found throughout history as well as in modern society. Military awards continue to be bestowed upon brave individuals who fight in the various wars of history. Mostly, these medals are given to recognize those who risked their lives to protect or aid their fellow soldiers.[31] We are moved and uplifted when we hear of such accounts of individuals self-sacrificing, even to the point of death, for their friends or fellow soldiers.

Often, the most heroic acts of altruism are performed during crisis moments. When a shooter opened fire in a classroom at the University of North Carolina at Charlotte, a twenty-one-year-old student named Riley Howell charged the shooter and knocked him off balance. This allowed for the shooter's capture by authorities and likely saved dozens of other students from death. Tragically, Howell was killed as a result of his heroic act, one of two fatalities from the shooting.[32] Lori Gilbert-Kaye, a sixty-year-old woman, performed a similarly fearless deed by jumping in front of a rabbi when a young man began shooting in her synagogue in San Diego. Gilbert-Kaye died from a bullet wound; the rabbi, who was trying to evacuate children when shielded by Gilbert-Kaye, survived with wounds in both of his hands.[33] Kendrick Castillo, an eighteen-year-old student at Highlands Ranch school in Colorado, lunged at a shooter in an effort to protect his friends and give them time to hide.[34] In these moments of crisis, these individuals, seemingly instinctively, sacrificed themselves to save others—touching examples of the deepest form of altruism.

And yet, as Edward O. Wilson notes, these incredible acts of altruism are merely "the extreme phenomenon that lies beyond the innumerable smaller impulses of courage and generosity that bind societies together."[35] Indeed, human society is built upon countless less heroic but still laudable acts of unselfishness. This drive to serve one another in unselfish ways has allowed us to become the most socially connected species on the planet; it is a deep part of our nature stemming from our evolutionary heritage.

A few decades ago, many biologists might have argued that this altruistic drive is culturally learned and not biologically programmed. Altruism, they thought, is imposed upon us by society and thus has little to do with biological evolution. This perspective dismisses any role of genes or biology in our tendencies to be kind. Obviously, culture and social learning can have a positive impact on altruism. Yet recent work has shown that we are also biologically hardwired to help others.

Over the last two decades, the psychologists Felix Warneken and Michael Tomasello devised clever experiments to observe altruism in babies. In one set of experiments, they tested the tendency of eighteen-month-old infants to help an adult they just met. There were several different "helping" tasks, where the adult feigned needing assistance in some form (dropping a marker, stacking books). In the majority of cases, the babies intuitively help the adult. And they would do it before he even made eye contact with the babies (a subtle form of asking for help). The babies simply discerned the intentions of the adult and were willing to help. In no case were the babies rewarded for any good behavior.[36] This experiment, or variations of it, has been repeated several times.[37] Even from a young age, before they have the ability to meaningfully communicate through verbal means, babies are primed to help others in need. What's more, as demonstrated in Warneken and Tomasello's experiment, the recipient of help doesn't need to be a family member. At least in some contexts, biology has clearly primed us to be altruistic.

◆

In the previous chapter, I argued that the deepest and most biological forms of altruism come from *kin selection*.* People who are closely related can be deeply motivated to act altruistically, even sacrifice for each other. Parents do this all the time for their children. The amount of sacrifice required to be a dedicated parent is astounding. In modern times, many parents essentially give up all their discretionary time and

---

* Different specialties define terms differently. Some biologists will say that *kin selection* only concerns the interactions between siblings (horizontal relationships). I am using the term *kin selection* more broadly (as might be used by anthropologists) and here intend for its meaning to include both horizontal (sibling relations) and vertical (parent-child) interactions.

live, as it were, through their children. This is (at least in part) because children carry their parents' genes. We are thus biologically primed to be altruistic toward (and sacrifice immensely for) our own children.

Yet kin selection is not the only way humans have developed altruism. As noted in the example of the *Grafton* survivors, people do not have to be biologically related to act altruistically toward each other, even when resources are scarce and immediate death is a real possibility. Friendship, altruism, and cooperation among nonkin are critical parts of our societies. How would kin selection account for this? Some might argue that friendship is just a spillover effect. We are biologically primed to be kind to kin, but our brains can direct this kindness beyond our actual blood relatives. There is likely some truth to this. As discussed in Chapter 3, even those with whom we are not considered blood relatives share 99.9 percent of our genes. Why squabble over the remaining 0.1 percent? There is a large amount of research confirming that becoming a parent helps to "civilize" adults, especially men. In essence, parenthood can direct us toward our altruistic and prosocial capacities.[38] For example, on average, men who are fathers and are invested in their children's lives show a greater degree of prosocial behavior than men who are not fathers or are not invested in rearing their children. One longitudinal study of parenthood among men found that masculine identities center around themes of aggressiveness and self-interest in the years prior to fatherhood. However, in the years following the beginning of fatherhood, their identities have shifted to center around their children as well as elements of caring and empathy, driving them to be less selfish.[39] This is because parenthood—through kin selection—has primed them to act altruistically, and this altruism can also be directed toward nonkin.*

* This concept will be discussed in more detail in Chapter 10.

Reciprocal Altruism and TIT for TAT

But other theories explaining altruism are also worth considering. Among them is reciprocal altruism, also known as direct reciprocity. Reciprocal altruism, as its name implies, is a sort of quid pro quo. Helping others and hoping, perhaps even expecting, that the recipients of your help will help you in return at some future date. The biologist Robert Trivers, who developed this concept, remarks that reciprocal altruism "take[s] the altruism out of altruism,"[40] a somewhat cynical perspective of human prosocial behavior. This concept takes advantage of the fact that, in social groups, interactions between people happen repeatedly over time. When small groups behave altruistically toward each other, trust develops and individuals are able to depend on this mutual altruism.

In 1950, two scientists[41] developed an imaginary game called the Prisoner's Dilemma that illustrates how reciprocal altruism may have come to be beneficial. To understand the game, imagine that two members of a gang are arrested and temporarily held captive. While being held captive, they are not permitted to communicate with each other in any way. Each prisoner is given a choice to remain silent and COOPERATE (with each other, not with the police) or to DEFECT and provide evidence against his partner (i.e., rat out the other). If both prisoners COOPERATE (by remaining silent), each will serve one year in prison. If both DEFECT, each will serve two years in prison. If the first DEFECTS and the second COOPERATES, the defector gets off scot-free (hence the temptation to DEFECT) but the cooperator gets three years in prison.[42]

If you were a participant in this imaginary game, the best outcome is clearly for you to DEFECT and your partner in crime to COOPERATE. But you have no idea what hand your partner will play. You can't communicate with him. If this social dilemma occurs just once, a large part of your fate is at the mercy of your partner's decision. But suppose this type of scenario happens over and over again. After all, in social groups, when individuals compete over resources and are constantly deciding whether to act selfishly or altruistically, these sorts of exchanges do occur repeatedly. This is what is called the *iterated* Prisoner's Dilemma. In 1980, a contest was held for budding game theorists and social scientists (or whoever else wanted to try their hand) for the best strategy to win in repeated rounds of the Prisoner's Dilemma. There were many responses, some of them extremely complex in their algorithms. But the winning strategy was actually the simplest (comprised of only a few lines of programming code). It was called TIT for TAT. This strategy started out by COOPERATING (playing the altruistic hand) on the first round. After this, however, its move copied (as the name would suggest) the previous move of its fellow prisoner. If the fellow prisoner DEFECTED, the TIT for TAT strategy would DEFECT on the next move. If the fellow prisoner COOPERATED, the TIT for TAT strategy would COOPERATE on the next move. Those strategies submitted to the contest that were selfish did very poorly in the long run. Those strategies that were altruistic tended to do better.[43] The results of the Prisoner's Dilemma tournament demonstrate, at least in an overly simplistic model, that reciprocal altruism may have allowed for humans to develop a form—albeit a superficial one—of altruism toward nonkin. Quid pro quo types of social exchanges abound in modern society.[44]

Indirect Reciprocity and the Importance of Reputation

Yet humans are astonishingly kind and altruistic even when it is anticipated that social exchanges will not be repeated, or when the giving can only go in one direction. There are plenty of examples of one-shot, anonymous giving in modern society: donating blood, giving money to starving children in developing countries, leaving a tip for a waiter (despite the fact that you will likely never see him again). Why do we do this?

Scientists have developed a theory of *indirect reciprocity* to help explain this sort of behavior. *Indirect reciprocity* has to do with our reputation.*[45] Put simply, the way we behave is observed by others. If I develop a reputation as someone who is altruistic, I am more likely to be the recipient of altruism, because others will see me as likely to return the favor. If we think of altruistic acts as a form of social currency (see Banker's Paradox, page 88), we can consider people who are altruistic as having higher "credit scores" than people who are selfish. In times of need, those who have developed a reputation of giving may be more likely to be helped than those who have earned a reputation as a miser.

Support for this hypothesis comes from finding that when our behavior is observed, we tend to act in more altruistic and cooperative ways.[46] This is a notion that anyone will readily agree is true. Think back to grade school to confirm this. If a teacher is present in the room, the class bully generally keeps his behavior in check. But once the teacher leaves, the behavior of children can quickly become problematic. A number of social science experiments confirm this intuitive aspect of human nature: people behave better when their behavior is observed.[47]

* Indirect reciprocity is sometimes called the reputation hypothesis. This was developed in part by the biologist Richard Alexander. Alexander noticed that the way hunter-gatherers shared food did not follow a simple tit-for-tat model (as Trivers might have predicted), but was based more on having a reputation of being generous and altruistic. In other words, "If I'm generous to someone today, *someone* will be generous to me in my time of need."

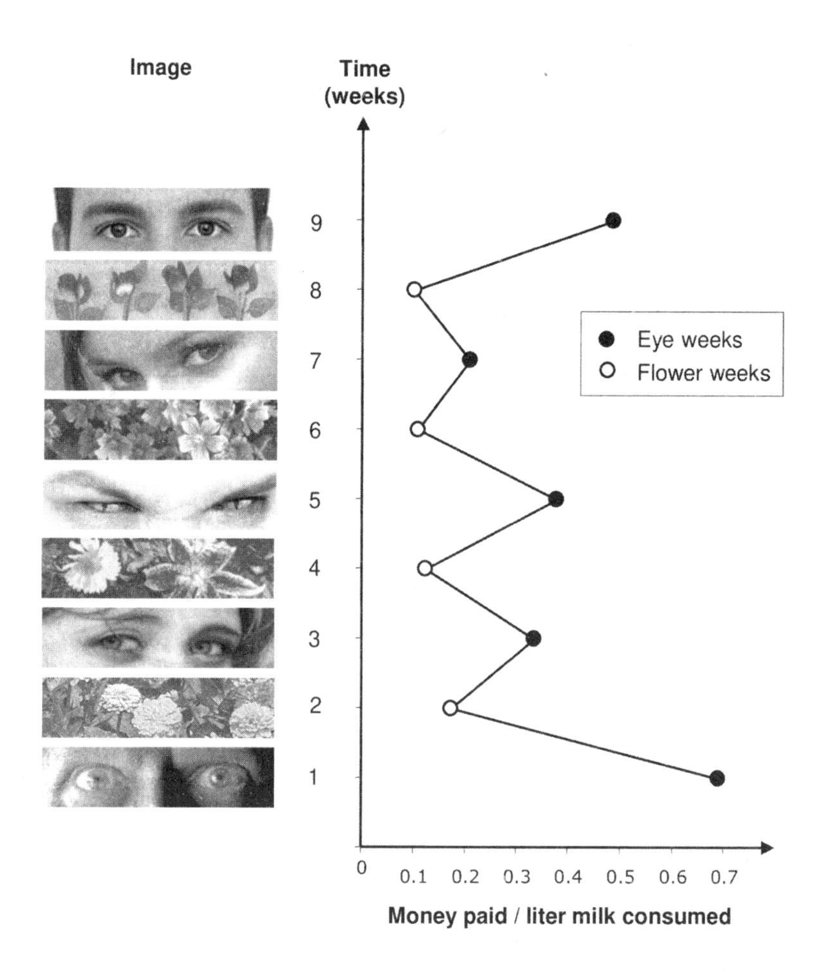

Figure 4.1. Amount of money voluntarily paid per quantity of milk consumed based on the "watchdog" image type. Adapted and reprinted with permission, from Bateson, et al., 2006.

Fascinatingly, the observer does not even have to be human, or even an animate object. The researcher Melissa Bateson and her colleagues devised an ingenious experiment showing that a coffee mug with an image of eyes on it could do a fair job as a watchdog of prosocial behavior. Bateson measured contributions to a box used to collect money for coffee that was shared in the commons area of a university dormitory. She situated a coffee mug at eye level, near the area where coffee drinkers are supposed to give to the collective fund. Every week, the picture on the coffee mug would alternate. On the odd weeks, there was a picture of human eyes, looking in the direction that a would-be coffee drinker would be positioned. On the even weeks, it was a mug with a picture of flowers. During the weeks when eyes were watching, students gave three times the amount to the collective coffee fund compared to when the mug displayed a picture of flowers (see Figure 4.1).[48] Indirect reciprocity and the effect of reputation can have a powerful effect on human behavior.*[49]

The High Evolutionary Value of Human Relationships

Yet direct and indirect reciprocity cannot fully explain the depth of human relationships. Humans readily give in anonymous ways. This clearly violates the assumptions of both direct and indirect reciprocity. Consider also friendship. A true friend does not depend on a quid pro quo exchange, nor does she develop or maintain a friendship to enhance her reputation. A good friend is someone who is there for you, despite the fact that you have nothing to offer. Friendship is not merely a transactional relationship. In fact, expecting some sort of favor in return is a telltale mark of *not* being a true friend.[50] We sometimes refer to this sort of superficial arrangement as fair-weather friendship. In contrast, humans find deep friendship rewarding because of the value of the relationship itself. There is nothing more emotionally satisfying than to know that someone cares about and loves you, even

* Another intriguing experiment: If you put a picture of eyes near a bus stop, people will pick up more litter than when there are pictures of flowers.

if you can't offer them anything in return except the same caring and compassion. At some point in our evolutionary past, deep relationships became extremely important for survival. As a consequence, they also became extremely rewarding. In other words, evolution seems to have assigned very high value to human relationships. Our ancestors faced a plethora of unpredictable challenges, including injury, illness, bad luck in hunting or gathering, and harsh weather.[51] The ability to develop deep relationships, where humans looked after one another in bad times as well as good, was crucial. In fact, these deep relationships became as important for survival as those resources we typically associate with survival in a Darwinian sense, namely food, shelter, and sex. As such, these deep relationships also became as rewarding as (and at some point, even more so than) these traditional resources. Forging committed relationships with kin as well as nonkin became essential not only for material survival, but also for emotional well-being. This is, at least in part, due to the high evolutionary value of relationships.

The evolutionary value of relationships helps solve what the psychologists John Tooby and Leda Cosmides coined as the Banker's Paradox. This paradox expresses the puzzle of the evolutionary origins of friendship (especially with nonkin) and our species' collective tendency to help those in need, especially when those in need are the least likely to repay us. Tooby and Cosmides describe the paradox this way:

> Bankers have a limited amount of money, and must choose who to invest it in. Each choice is a gamble: taken together, they must ultimately yield a net profit, or the banker will go out of business. This set of incentives leads to a common complaint about the banking system: that bankers will only loan money to individuals who do not need it. The harsh irony of the Banker's Paradox is this: just when individuals need

> money most desperately, they are also the poorest credit risks and, therefore, the least likely to be selected to receive a loan.[52]

Similarly, people have limited social resources. Our ancestors had to choose in whom to invest their social capital. Considering only the perspective of reciprocal altruism, the person in greatest need is also the highest "credit risk," or least likely to be in a position to return the favor. Reciprocal altruism would not favor the evolution of an inclination to help and befriend the helpless. Yet all societies have some form of social mechanism to help the needy. The development of deep relationships where friends are there for you in your most desperate hour was nature's solution to the Banker's Paradox.[53]

This capacity for forming relationships that are not based on exchange of goods, services, favors, or on reputation has at least some of its roots in kin selection. Parents have developed the capacity to have deep relationships with their children. Compared with other species, human offspring mature at an extremely slow rate. They are dependent on their parents for material and emotional support for the better part of fifteen years. As infants, they are utterly helpless and cannot do anything at all for themselves. It stands to reason that humans developed a capacity for deep compassion for their children to allow for their nurture and development. Many people would willingly undertake considerable personal effort to help a young mother or her new child, even if they are not direct relatives. In the vast majority of cultures, people have developed a natural inclination to be protective of and altruistic toward babies, young children, and their mothers. Given this very long maturation period of human children, kin selection is perhaps the first way our inclination to help the helpless evolved.

Kin selection predicts that parents behave altruistically toward their children. In humans, because of the extremely long maturation process of our children and their utter helplessness at birth, kin selection must have made it extremely rewarding to develop a deep and enjoyable relationship with children. Even today, when children are sometimes viewed as enormous financial liabilities, many people routinely choose to have children. Why? Because parenthood is deeply satisfying (at least it has the capacity to be), despite the fact that it is also incredibly challenging. Evolution has biologically primed us to care deeply for our children *and to enjoy caring for them*. This is how evolution created the most powerful forms of love. A purely technical approach that overlooks the evolutionary value of human relationships and the emotional aspects of kin selection runs the risk of missing this point. The same could be said of reciprocal altruism, or *tit* for *tat*. Not only do we benefit in a material sense from being the recipient of an altruistic act, but evolution has also shaped us so that we often benefit emotionally from being the giver of such an act.

A plethora of psychological experiments has confirmed the rewarding nature of altruism. In one study, researchers gave university students some cash in an envelope. Half of the students were instructed to spend the money on themselves—either on a bill, an expense, or a personal gift—by the end of the day. The other half were told to spend it on someone else or donate it to a charity. Surprising to most people (including the students themselves), those who spent the money on others were happier by the end of the day.[54] This experiment, or variations on it, has been repeated many times, with the consistent finding that altruism is rewarding. In children as young as twenty-two months, toddlers who give away treats are happier than those who receive treats. In fact, more costly giving seems to lead to more happiness than non-costly giving. In other words, giving away something that is yours (requiring a level of sacrifice) as opposed to giving away something that isn't yours, brings more emotional reward.[55] In another study, small children in rural communities were asked to share either their own candy (costly giving, because it required a sacrifice) or share an additional piece of candy (non-costly giving, because it was extra). Not only did costly giving result in more happiness compared to non-costly giving, but the children exhibited more happiness when giving away candy compared to receiving candy.[56] Perhaps it really is true, as the saying goes, that it is more blessed to give than to receive.[57] What's more, the emotional benefits of altruism are apparent even when there is little or no contact with the recipient, suggesting that the rewarding nature of giving is not simply a result of anticipating social praise or reciprocity.[58]

What is going on here? These findings seem to fly in the face of our traditional conception of evolution and a survival of the fittest. Why would evolution mold us to feel that giving for the sake of giving is rewarding, especially to nonkin? Again, the answer likely has to do with the high value human evolution has assigned to deep relationships. The rewarding nature of altruism for the sake of altruism is just one piece of evidence that relationships were extremely important in our evolution.

Another body of evidence for the high evolutionary value of relationships is that having caring relationships is one of the strongest predictors of human health, happiness, and well-being. There is an immense amount of psychological data to support this.* Over and over again, one of the key takeaways from the psychological sciences is that deep and committed relationships are more powerful in predicting happiness than wealth, social status, education, or a host of other factors.[59]

Consider what happens when these deep relationships are lacking. Today, our society has created systems (i.e., social media) in which we come into contact at breathtaking speed (from an evolutionary

---

* I discuss this concept in more detail in Chapter 9.

perspective) with a huge number of people. But most of these social interactions are shallow and hence not emotionally fulfilling. As humans, we are extremely prone to being lonely. The COVID-19 pandemic reminded us all too painfully of this aspect of human nature. Loneliness can heighten the risk of depression, anxiety, and even suicidal thinking.[60] With so many of our interactions occurring on digital platforms, we are increasingly immersed in superficial social exchanges. In a hyperconnected world, if we lack deep relationships with friends and family, we experience social starvation in the midst of plenty.

Even today, when some might say that the strong biological forces of evolution have given way to cultural forces, strong relationships predict longevity and protect against death. In fact, loneliness (the opposite of having strong relationships) seems to be as harmful to physical health as obesity or cigarette smoking.[61] If this is the case today, imagine how important strong relationships were when the climate was exceptionally volatile and environmental pressures for survival were extreme (as scientists affirm was likely the case in our recent evolutionary past). The collective altruism and cooperative spirit fostered by the crew of the *Grafton* help us imagine scenarios in which strong relationships were critical for survival in our evolutionary past.

In sum, the high importance of relationships in the survival and thriving of human social groups helps us understand many things that might not initially fit with Darwin's theory of evolution. It also helps us understand why, in modern society, we have developed deep and personal relationships with nonkin. This is what we call friendship. Whether friendship also has roots in group selection or is a spillover effect from kin selection (or has some other origin) is debatable. However, that evolution favored deep relationships among human kin—partly because of the extremely long maturation period of infants and their utter helplessness at birth—is beyond dispute. As parents, we are biologically primed to love and care for our children.

◆

Natural selection operating on an individual level has clearly primed us to be capable of great selfishness. This was the traditional Darwinian picture of nature that evolution painted. Life was depicted as a state of constant struggle against each other in which only the strong and selfish survive. Certainly, humans have a deep capacity for selfishness. Yet curiously, nature has also endowed us with the capacity for immense altruism. Somewhat paradoxically, these tendencies for altruism seem to be just as inborn as those for selfishness. Evolutionary mechanisms such as kin selection and direct/indirect reciprocity have built within us a nature seemingly in opposition to the one created by an individual-level selection.[62] Edward O. Wilson has observed how, in humans, these evolutionary forces predict a sort of contradiction and tension as a way of life: "No wonder the human spirit is in constant turmoil."[63] Evolution has essentially pitted these biological and social drives against each other.

Fortunately, the evidence discussed above about friendship, altruism, and the way in which we crave deep and meaningful relationships demonstrates that, between these two tendencies, altruism is the stronger force. Yet the pull of selfishness is strong and can threaten not only individual well-being but also whole societies. On an individual level, how many of us don't regret a selfish act that endangered a close relationship with a friend or family member? Our society's emphasis on economic and social status sometimes tempts us to devote countless hours to career development (which if we are not careful, can become a purely selfish pursuit), often to the detriment of our personal and family relationships.

Given the inherent tension in human nature between selfishness and altruism, our societies would do well to understand what environments, situations, and social arrangements are more likely to elicit the better angels of our nature. As the deepest forms of altruism derive from the evolutionary forces acting on family relationships, cultures that serve to strengthen the emotional ties between parents and offspring, as well as with extended family, would likely see immense overall social benefits.*

* This topic is explored in more depth in Chapter 10.

Chapter Five
Aggression and Cooperation

"Throughout . . . human evolution . . . there was . . . one pattern of social behavior that became extremely important, and we can therefore expect the forces of natural selection would have ensured its becoming deeply embedded in the human brain: this is cooperation."[1]

—Richard Leakey and Roger Lewin

"Human beings are not gentle creatures in need of love . . . on the contrary, they can count a powerful share of aggression among their instinctual endowments. Hence, their neighbor is . . . someone who tempts them to take out their aggression on him, to exploit his labor without recompense . . . to take possession of his goods, to humiliate him and cause him pain, to torture and kill him."[2]

—Sigmund Freud

In some ways, we are the most aggressive of all species. Human history is replete with a level of intraspecies violence not generally observed in the animal world. Individuals hurt, maim, and murder one another; nations fight bloody, destructive wars. No other species matches us in the cruelty we sometimes show to each other.

But we have also developed a remarkable ability for tolerance, trust, and understanding. We are the most cooperative of all species. Tasks that we might consider very basic forms of cooperation—such as doing the wave in a football stadium[3]—are extraordinarily rare among animals. Our most remarkable collective technological achievements—heart transplants, space travel, satellite communication networks, supercomputers, and so forth—are unmatched in the animal world. All are only possible because of our deeply rooted cooperative capacities.

As is the case with selfishness and altruism, evolution has paradoxically built within us opposing capacities of aggression and cooperation. We are hardwired, it seems, for both aggression and peaceful cooperation. But how is that possible? In this chapter, I'll try to answer that question by exploring how and why nature has endowed us with these contradictory tendencies.

Let's start with aggression.

Depending on the scientific discipline, aggression can be categorized in many different ways. Ethologists (who study animal behavior) concern themselves with aggression that is defensive (as when bees defend their hive) or offensive (as when hyenas steal a carcass). Psychologists might concern themselves with the purpose of the aggressive act: accidental, instrumental, expressive, and so forth. Some social scientists might concern themselves with the organizational level on which aggression occurs: all-out warfare between nations, gang violence, or one-on-one homicide. Anthropologists, for their part, might concern themselves with the distinction between aggression that is hot (emotional) or cold (premeditated). In this book, I'll adopt a relatively simple definition of aggression as the competition against others for some critical resource, such as food or shelter.

Similar to selfishness, the capacity for aggression is easy to explain from an evolutionary perspective. In a traditional Darwinian model, the aggressive organism is able to outcompete others and so has a better chance of survival and reproduction. From a more narrowly biological standpoint, aggression serves the purpose of enabling an organism to compete for important but limited resources. Aggression in humans is no exception. Violence is a special case that arises when aggression escalates to the point of physical harm. Humans undeniably have the capacity to be violent: The terrible crimes of domestic violence, rape, and murder occur in virtually all cultures across the globe.

One of the main debates among anthropologists these days is how much of the root of human aggression comes from biology and how much from culture.[4] Are our aggressive tendencies more innate than learned?

In this debate, we might hope that aggression is more learned than innate, because learned behaviors seem to be modifiable whereas those rooted in biology do not. But that's a faulty assumption. Even though many aggressive behaviors are rooted in biology, it turns out they can still be modified. (Recall from Chapter 3 that the primary driver of whether genes are "on" or "off" is the environment.) Culture and environmental factors can affect biological systems to render species (including humans) more or less aggressive.[*5]

---

* For example, when new fathers become engaged in parenting their children, their levels of testosterone drop. There is something about the environment of caring for children (nurture) that leads to a biological change in the expression of testosterone (nature) in fathers. While the relationship between testosterone and aggression is complicated in humans, fathers who are engaged in rearing their children also show lower rates of aggression than their childless peers or fathers who are not engaged in parenting. This likely developed to help channel human male energy into the rearing of their own children.

Biological Factors

When we think of the relationship between aggression and biology, we almost always think of the hormone testosterone. For decades, many scientists believed that more testosterone equaled more aggression. To some degree and in some contexts, this simple rule holds. But in general, the truth—especially in humans—is much more complicated.

During pregnancy and adolescence in humans, testosterone plays an important role in determining the biological differences between males and females. It also performs roles important to all humans, including the maintenance of bone and muscle mass and holding the inflammatory system in check. In this respect, the difference between men and women is not that men have testosterone and women don't, but that men have it to a much greater degree. (One recent study suggests that men typically have about fifteen times more testosterone than women.[6])

The way in which testosterone and other hormones interact is extremely complicated, yet testosterone has been shown to dramatically influence aggression in many animals. The German scientist Arnold Berthold was one of the first to show this in 1849. Berthold observed drastic changes in the behavior and development of immature male chickens after removing their testes (the primary site of testosterone production). These castrated chickens—called capons—did not engage in typical territorial disputes with other roosters. They did not mate with females. They were much less aggressive than the typical rooster. However, when Berthold took testes from other roosters and surgically inserted them into the abdomens of capons, the capons would start acting like normal roosters (including showing aggression). They fought with other roosters, crowed, and sought opportunities to mate. When the transplanted testes were removed, the male chickens again behaved like capons.[7]

Berthold didn't know it at the time, but subsequent experiments showed that testosterone is the key driver behind this observation. Injecting testosterone (as opposed to implanting the entire testes) into castrated chickens produces the same results as Berthold's original experiment. Similar effects of testosterone have also been documented in a range of species, including some types of lizards, painted turtles, doves, songbirds, grouse, deer, rats, swordtail fish, quail, mice, and chimpanzees (to name just a few).[8] The notion that testosterone can exert powerful influence on aggressive behavior is beyond dispute.

A number of other biological factors can also play a role in aggression.[9] Recall the "warrior gene" from Chapter 3. The real name of this gene is the MAOA gene (short for monoamine oxidase, type A). The MAOA gene makes a protein that breaks down certain types of neurotransmitters that are part of the brain. One version of this gene (MAOA-L, where L is for low activity) is rather inefficient at breaking down these neurotransmitters. This disruption is thought to cause some people to have less emotional control, to take greater risks, and to be more aggressive. In one experiment, those carrying the MAOA-L gene were more likely to administer higher quantities of extremely hot chili sauce to those whom they perceived had wronged them.[10] Another study found a consistent but small link between men with the MAOA-L gene and the tendency to exhibit high rates of antisocial behavior.[11]

In addition to hormones and genes, brain-activity patterns have also been linked to aggression. In the 1990s, the psychologist Adrian Raine scanned the brains of forty-one men held in the California penitentiary system, all of whom were being tried for murder. He also scanned the brains of forty-one "healthy controls," men who were matched for age but who had not been accused of murder. Compared with the healthy controls, the men in prison had higher brain activity in what scientists call the subcortical and limbic parts of the brain (areas that help regulate emotions). This finding led Raine to conclude that, in general, killers tended to experience stronger emotions.[12] The abnormally high levels of activity of these brain areas may have contributed to the mental calculations that led these men to kill.

On first glance, it might be tempting to conclude that high testosterone, a certain type of warrior gene variant, or a hyperactive limbic system (or some other biological factor) leads to the inevitability of violent behavior. Taken to the extreme, this view is called biological determinism. In other words, our biological makeup (as opposed to culture, environment, or personal decisions) determines our behavior. In the case of testosterone, higher levels of the hormone have often been found to be associated with criminal behavior in many (though not all) studies.[13] Indeed, 91 percent of individuals who are incarcerated in the United States are men.[14] The pattern is similar with violent crime. Based on global statistics compiled by the United Nations in 2019, men account for over 90 percent of perpetrators—and 81 percent of victims—of homicide.[15]

This notion of biological determinism was an approach used in 1994 by the attorneys of Stephen Mobley, who had been accused of homicide in the death of a young pizza-shop manager. The evidence was overwhelming, and Mobley had even confessed several times. In an effort to reduce his sentence from capital punishment to life imprisonment, his attorneys offered a defense based on biological determinism. Mobley, they argued, came from a long line of criminals. His criminal behavior was genetic and therefore rooted in biology. Upon analyzing several generations of Mobley's family tree, it was discovered that individuals from at least four generations had committed acts of violence and aggression that included rape, murder, explosive tempers, alcoholism, and spousal abuse.[16] His attorneys then cited a recently published research report in the prestigious journal *Science*, citing evidence that the MAOA-L gene variant (the "warrior gene") is linked to aggressive behavior.[17]

But the fact that biology plays a key role in human behavior does not make aggression inevitable. Our behavior is not biologically predetermined. (The judges who tried the Mobley case seemed to agree; the motion to accept scientific evidence diminishing Mobley's culpability in this gross act of violence was dismissed.) Humans are much more complex and variable than castrated chickens. Injecting normal humans with testosterone in an effort to achieve higher levels of the hormone does not generally make them more aggressive.

This is true even in such instances where men have abnormally low levels of testosterone, as is seen in a disorder known as hypogonadism. In male hypogonadism, the testes do not produce normal levels of testosterone. Men with hypogonadism are like human capons. When testosterone is used as a clinical treatment for these men, aggressive behavior is extremely rare as a side effect.[18] Furthermore, the relationship between levels of testosterone and various measures of human aggression is rather weak.[19] It may be true that those who commit violent crimes have higher levels of testosterone than average; however, it is also true that, among those who have higher than average levels of testosterone, the overwhelming majority do *not* commit violent crimes.[20] In fact, there is growing evidence that engaging in aggressive behavior can lead to an increase in testosterone—quite a role reversal of the cause and effect pattern we are accustomed to thinking about when it comes to testosterone and aggression. In other words, part of the reason that violent criminals have high levels of testosterone is because engaging in criminal activity likely leads to an increase in the amount of circulating testosterone. What's more, in some contexts, testosterone can rise or fall based simply on a person's attitude or expectation of their performance.[21]

Nature and Nurture

There is no gene, nor any set of genes that we know of, that leads to aggressive behavior. Some individuals are more prone to aggression than others, whether by virtue of biological or environmental factors or a combination of both, but biology is not fate. When considered alone, in fact, genetic differences and other biological factors only weakly predict aggressive behavior.* As the Harvard anthropologist Richard Wrangham, an expert on aggression, has put it, genes can "influence [aggressive] behavior" but "rarely determine it."[22]

Cultural and environmental factors both play an important role. One of the first studies providing compelling evidence of the cultural contribution to human aggression came from the anthropologist Richard Sipes in the 1970s. At the time, the prevailing view was that human aggression could be understood using a "drive-discharge" model. Like a hydraulic fluid that builds up over time, sooner or later aggression must find an outlet. This outlet could come in many forms, the most extreme being all-out warfare. However, the theory went, perhaps societies could find ways for boys and men to discharge their aggressive drives in non-mortal ways. Combative sports[23] (like American football or boxing) might serve as alternative outlets to war. If the "drive-discharge" model was correct, we would expect to find

* The key exception to this is the genetic determinant of sex: the Y chromosome. Men have been shown to be consistently more aggressive than women.

fewer wars among societies with combative sports. In other words, if aggressive behavior is predominantly biological, there should be comparatively equal levels of aggression across cultures, regardless of whether this aggression is expressed in the form of war or combative sports.

Sipes sampled societies from across the world to test this theory. What he found strongly refuted the notion that aggression is like a drive that needs to be discharged. Among the most warlike societies, 90 percent had combative sports. On the other hand, combative sports were only present in 20 percent of the most peaceful societies. Sipes concluded that aggressive behavior could produce a sort of kindling effect, much like fanning the flames of a fire. Indeed, the presence of war seemed to encourage greater levels of other aggressive activities in societies. His study contradicted the "drive-discharge" model of human aggression in favor of a "culture-pattern" model of human aggression. Aggressive behavior, he posited, is primarily learned.[24]

Sipes's observation that aggression could be a learned behavior was in line with the work of the Stanford psychologist Albert Bandura. In the 1960s, Bandura and his team conducted a series of experiments where small children watched how adults behaved toward a Bobo doll (a doll-like toy with a rounded bottom and low center of mass such that it would pop back up if knocked down). Half of the children watched adults acting aggressively toward the doll for ten minutes (kicking, hitting, or punching the doll). The other children watched adults who merely played with other toys and ignored the doll for ten minutes. The children were then observed as they were left to play by themselves. Anyone who is a parent can guess what happened next. The children who watched adults behaving aggressively exhibited high rates of physical and verbal aggression; the children exposed to the nonviolent scenes exhibited almost no aggression.[25]

Over the past several decades, social scientists have amassed a great deal of data showing how aggressive behavior can be learned,*[26] in line with the findings of Sipes and Bandura. In fact, some researchers now liken the spread of violence through social learning to a contagious disease.[27] Individuals, especially children and adolescents, can "catch" the violence bug merely from being exposed to others who are violent. One surprising finding from this new research is that aggressive learning can even travel upstream; not only can children become violent from the influence of their parents, but parents can become violent from the influence of their children.[28]

The University of Michigan professor Rowell Huesmann and his team collected data for several years from children who were suffering through the ongoing ethnic-political violence of the Israeli-Palestinian conflict. They interviewed over 1,200 children and parents several times over a three-year period. The rates of exposure to violence were appalling: 55 percent of children had seen a friend die, 43 percent had seen someone tortured or held hostage, and 63 percent observed someone crying because someone they knew had died.[29]

One key question that Huesmann and his team were trying to answer was: Does exposure to violence increase the probability that children and adolescents will be aggressive and violent toward their peers? The answer was definitively yes. Those who were exposed to the

---

* Another piece of evidence of the cultural contribution to aggression comes from the widely varying rates of violence found in different countries. Carole Hooven gives one instructive example: "The rate of violent crime in Singapore is minuscule compared to Jamaica, the United States, or even relatively peaceful Canada, whose assault rate is about fifty times greater than Singapore's. Singapore has the lowest murder rate in the world, alongside Japan. Why is Singapore so different? The government doesn't pump pacifying chemicals into the water supply. The explanation presumably lies in the Singaporean culture, which is one of law-abidingness, strict discipline in families, and lack of poverty, plus harsh criminal penalties, among other factors."

most violence (the top 25 percent) were twelve times more likely to choke a peer, twice as likely to knife (or threaten to knife) a peer, and one-and-a-half times more likely to punch or beat a peer compared to those who were exposed to the least violence (the bottom 25 percent).[30]

Of course, not all children who were exposed to violence in the Israeli-Palestinian conflict became violent themselves. Why some did and some didn't is still an open question. Biological factors likely play a role. Hence, as the Harvard biologist Edward Wilson states: "It is true that aggressive behavior, especially in its more dangerous forms . . . is learned. But the learning is prepared. . . . We are strongly predisposed to slide into deep . . . hostility under certain . . . conditions. With dangerous ease, hostility feeds on itself and ignites runaway reactions that can swiftly progress to . . . violence."[31] Unfortunately, the capacity for aggression and even violence is a part of human nature, but both biology and culture play important roles in its genesis.[*32]

Premediated and Impulsive Aggression

How do humans compare to animals with respect to aggression? It depends. Broadly speaking, aggression in humans can be categorized into two different types. Compared to most animals, our aggression is less reflexive and more premeditated. This dichotomy is known by several names: hot versus cold, reactive versus proactive, impulsive versus premeditated, reflexive versus deliberate. For our purposes, I'll use the terms impulsive and premeditated. Compared to animals, especially

* One of the best and most succinct phrases I have read that describes the interplay between biology and culture with respect to many facets of human nature comes from Richard A. Lippa, who writes that "cultural influences [are] superimposed on biological predispositions."

those that scientists tell us are our closest biological relatives, humans have a very low level of impulsive aggression but a high capacity for premeditated aggression.

The anthropologist Sarah Hrdy offers a classic example of the unique balance of these two types of aggression in humans. Suppose, Hdry imagines, you were to try to board an airplane with 200 chimpanzees. The result would be violent chaos. You would be lucky, she notes, "to disembark with all ten fingers and toes still attached." And she continues, "Bloody earlobes and other appendages would litter the aisles. Compressing so many highly impulsive strangers into a tight space would be a recipe for mayhem."[33] Yet such an occurrence happens thousands of times per day in human societies, with very few instances of this type of impulsive violence.

However, the boarding of an airplane is preceded by an intense security and screening procedure. This process is designed precisely to prevent less impulsive but more deadly forms of aggression by would-be hijackers and terrorists. Terrible examples of this type of aggression come all too quickly to mind. And then there's war, the deadliest form of this type of aggression, which in virtually all circumstances is an expression of premeditated (as opposed to impulsive) aggression.

Warfare and terrorist acts are just extreme instances of premeditated aggression. Less intense examples abound in modern society. Gossiping, bullying, even aggressive business practices can be forms of premeditated aggression. Unlike impulsive aggression, premeditated aggression is not carried out in the heat of the moment. It requires planning, coordination, and a level of intelligence that is not needed for impulsive aggression.

To be sure, humans can still be provoked to impulsive aggression. Virtually every human being has at one point lost his or her temper, been provoked to anger, or yelled at a loved one only to later regret it. Even more dangerous forms of impulsive aggression are far too common. Domestic violence, for example, is largely a problem of impulsive aggression. The World Health Organization estimates that the proportion of women worldwide who report having experienced physical or sexual violence at the hands of their partner is 26 percent.[34] The capacity for impulsive aggression remains an unfortunate part of our natures. As a society, we should not be satisfied with any rate of domestic violence above zero. However, our capacity for impulsive aggression is much lower compared to our closest biological relatives: among chimpanzees, 100 percent of females experience domestic violence.[35] *One hundred percent!*

If humans were as impulsively aggressive as chimpanzees, we would not be able to carry out the complex and cooperative tasks that are critical to society. So why are we so much less impulsively aggressive than they are?

Richard Wrangham and others have argued that we've achieved this reduction in impulsive aggression by domesticating ourselves, just as we've domesticated dogs, sheep, goats, chickens, and other farm animals.*[36] Although this process appears to be driven largely by culture, it has certainly affected our biology. In human evolution, mating preferences were thought to have played an important role in self-domestication. So, too, did the efforts of primitive social groups

* There are several examples of species that seem to have been domesticated over many generations and thus began to diverge from their original species in form and behavior. This is similar to the way that anthropologists think we diverged from our ancestors. For example, dogs are considered a domesticated form of wolves. It may have been evolutionarily advantageous for wolves to domesticate into dogs so that they could develop a mutually beneficial relationship with humans, scavenging off of remnants of food and such. Over time, the wolf became less and less aggressive and less anxious around humans.

to guard against an overly aggressive leader or alpha male. There is evidence suggesting that self-domestication indeed happened to some degree in humans, and that it led to our species' relatively low levels of impulsive aggression when compared to chimpanzees and other great apes.[37] For example, among the !Kung people native to Botswana, it appears that the most common reason for verbal criticism between members of the clan was to temper "big-shot" behavior found in arrogant males. There is even reason to believe this was also the case in human cultures in Africa dating back to 45,000 years ago.[38]

Why Are We Cooperative?

Notwithstanding our species' capacity for aggression and violence, we are remarkably adept at working together. More than any other primate, humans form extensive social groups and perform complex tasks that require collaboration with others. How did we become so cooperative? What is the evolutionary origin of this trait?

Initially, biologists believed that reciprocal altruism explained the development of human cooperation. This relates back to the TIT for TAT algorithm developed for the iterated Prisoner's Dilemma. If you recall, this is a simplified situation of repeated social exchanges where two players each have the opportunity to COOPERATE or DEFECT without knowing what the other will do. It is in the interest of both players to cooperate over the long term, yet if one player DEFECTS at the same time another COOPERATES, the defector will reap a larger short-term reward. Strategies that encourage cooperation fare the best, while those favoring selfishness perform poorly. In some ways, the TIT for TAT explanation of cooperation is powerful. This seems especially true among our closest relatives, chimpanzees.*[39]

Yet in other ways, the TIT for TAT strategy is a superficial explanation of human cooperation (just like it is a superficial explanation of human altruism). Although examples of cooperation based on a quid pro quo abound in modern society, subsequent studies of human nature suggest that our ability to cooperate is much deeper than a simple TIT for TAT exchange. Psychological experiments conducted with children—including those conducted by Felix Tomasello and Michael Warneken, as discussed in Chapter 4—have shown that we have a natural drive to cooperate. In most of their experiments, Tomasello and Warneken found that young children cooperate immediately (without being asked), that they will help strangers, and they will do so without the promise of a reward.[40] When the researchers introduce the promise of a reward, it doesn't influence the children's cooperative behavior. In fact, those children who received rewards were less likely to cooperate in subsequent experiments.[41] This observation (known to psychologists as an *overjustification effect*) suggests that cooperation itself is intrinsically rewarding. When an external reward is provided, it can highjack this internal drive and paradoxically weaken it.

Tomasello has worked for decades on this topic and has conducted dozens of experiments—all of which have led him to conclude that our drive to cooperate is a natural predisposition.[42] He concludes this for at least four main reasons: 1) Cooperation occurs so readily in children before they have the ability to speak and before their parents explicitly teach them to be helpful, 2) it occurs in infants across a variety of

---

* Chimpanzees show forms of cooperation, but these are mostly limited to close kin and reciprocating partners. Humans, however, show a much more pervasive form of cooperation and are willing to incur costs to help even strangers in anonymous one-shot interactions.

cultures, 3) it occurs without the promise or expectation of a reward or verbal praise from a parent, and 4) we can see rudimentary forms of it in chimpanzees, our closest evolutionary relatives.[43]

Our innate drive to cooperate, even from infancy, could not have come entirely from reciprocal altruism (i.e., the COOPERATE and DEFECT scenario from the repeated Prisoner's Dilemma experiment). Then where did it come from? How are we so cooperative, even when we have the capacity for premeditated aggression?

Group Dynamics

Some scientists have argued that these opposing capacities are actually related and developed jointly in our evolutionary history. Evidence for this comes in part from observations of modern hunter-gatherer societies that anthropologists tell us are similar to our ancestors in behavior and group dynamics.

In 1961, for example, the anthropologist Karl Heider lived for a time among the Dani farmers in the beautiful Grand Valley of New Guinea. He later reported his observations and experiences in a book, subtitled *Peaceful Warriors*, in which he described the Dani as a people of gentle demeanor who were slow to anger. They lived in mutually dependent and supportive social systems, and as long as they had no conflicts with other tribes, they were a calm and pacific rural people. But when tribal conflict arose, the Dani became unspeakably violent.[44] As Richard Wrangham has written, summing up Heider's observations: "The Dani had one of the highest killing rates ever recorded. . . . If the rest of the world had been like the Dani, the 20th century's sickening 100 million war deaths would have ballooned to an unthinkable 2 billion."[45]

Many other examples can be found in the anthropological record. Among the Dayak, in Borneo, a man must slaughter an enemy before he is allowed to marry, yet killing within the tribe is a terrible offense.[46] Similarly, the Dayak tolerate and even encourage killing members of other tribes but absolutely forbid it within the tribe. As one anthropologist observed, writing in the early 1900s, "Not only is slaying an enemy in open war looked on as righteous, but ancient law goes on the doctrine that slaying one's own tribesman and slaying a foreigner are crimes of quite different order."[47] In northeast India, the Angami people are generally reported to be peaceful and honest among themselves. But toward foreigners, they are "bloodthirsty, treacherous, and revengeful to an almost incredible degree."[48] Similar patterns of intragroup peace and intergroup hostility have been seen in the Native American communities that were in existence when Europeans began colonizing North and South America.

In all of these cases, the pattern seems clear: people in small-scale, hunter-gatherer societies are as good to insiders as they are harsh to outsiders.[49] Psychologists refer to these phenomena as *in-group bias* and *out-group bias*. Such biases are the basis of the tribalism and xenophobia that we find in larger societies all over the world today. We are kind to those we identify as part of our group, but cold (or even hostile) toward those who are not.

But what constitutes a group? People with the same skin color? Political affiliation? History and experience would suggest that, unfortunately, these distinctions can indeed set the stage for powerful in-group benevolence and out-group animosity. Over the years, psychologists have shown through numerous experiments that humans can be made to identify with groups based on even the subtlest distinctions.[50] In one social experiment, a researcher disguised as a jogger pretended to trip and hurt his ankle in front of soccer fans en route to watch their favorite team play a game. How many of the soccer fans offered to help the fallen jogger? It depended on if he was wearing the in-group team's shirt versus a rival team's shirt.[51] Even young children, notwithstanding their innate cooperative natures, show a preference for other children who are randomly assigned the same color of T-shirt.[52]

The fact that humans seem to innately identify with and show favor for someone who is part of their social group helps explain why some scientists believe that cooperation and aggression evolved together. A classic social experiment, known as the Robbers Cave experiment, powerfully exhibits how this might be the case.[53]

In 1954, the psychologist Muzafer Sherif and colleagues invited a group of 22 twelve-year-old boys to a special summer camp at Robbers Cave State Park in the rural mountains of southeastern Oklahoma. The camp had all the fixings for the making of a classic summer camp experience for young boys. These boys had never before interacted but came from similar backgrounds. They were socially well adjusted, academically successful, and hailed from stable two-parent families.

As the boys arrived, the researchers, posing as camp personnel, randomly divided the youngsters into two groups of eleven. For the first week of camp, each group was placed far enough away from the other group that there was no direct interaction between the groups' members. In fact, initially, each boy thought there were only eleven boys to begin with. During this first week, the experimenters observed how quickly group identity could be formed among strangers. The boys were allowed to bond through swimming, canoeing, and preparing meals together. They were assigned activities and tasks that would appeal to most young boys but that were designed to compel them to cooperate, plan, and execute tasks as a group (i.e., carrying canoes to the lake). After a week, each group formed its own distinct identity. Leaders emerged and social norms were established, with sanctions for deviant behavior. The boys chose names for their groups—the Eagles and the Rattlers—and even stenciled these team names onto shirts and flags.

During the second week of the experiment, Sherif and his research assistants examined how the social dynamics would shift when these newly formed groups were placed into competition with each other. The experimenters set up tournaments between the Eagles and the Rattlers. These competitions were carefully chosen to be zero-sum games: there always had to be a winner and a loser, so as to foster maximum between-group competition (i.e., tug-of-war, touch football, etc.). To up the ante, the camp directors announced that prizes would be awarded for the winners of the competitions (pocket knives, medals, and trophies).

Expectedly, the groups soon developed negative attitudes toward each other. Even physical aggression arose, forcing the experimenters to intervene from time to time. The Eagles burned the Rattlers' flag, which prompted the Rattlers to retaliate by ransacking the Eagles' cabin in the middle of the night. The Eagles retaliated the following morning, scattering dirt and possessions throughout the Rattlers' campsite. At the same time that conflict was developing between the groups, solidarity and cooperation were increasing within them. The boys were becoming increasingly aggressive toward those of the opposing group while becoming more cooperative with those from their own group. The whole experiment was a classic case of in-group and out-group dynamics made to play out with rapid timing.[54]

Mathematical modeling has also supported the notion that cooperation and aggression may have coevolved. The economist Samuel Bowles, who has developed many of these mathematical models, even refers to intergroup conflict as "altruism's midwife." Bowles explains that he and his colleagues have developed simulations that approximate the thousands of generations that interacted as our ancestors evolved. Groups are composed of individuals with varying degrees of altruism and selfishness, along with varying gradations of tolerance or hostility toward those outside the group. Taking into account the extreme climate instability that scientists think occurred while our ancestors were developing, Bowles has found that the development of altruism and warlike tendencies might be dependent upon one another. His models suggest that neither altruism (a key ingredient for cooperation) nor the capacity for warlike behavior would evolve alone. Each capacity needs the other.[55] Remarking on this somewhat unexpected finding, the Yale sociologist Nicholas Christakis notes: "It's a puzzle that humans can be so friendly and kind but also so hateful and violent. . . . Maybe, therefore, kindness and hatred are actually related. . . . In the past, conditions were ripe for the emergence of both altruism and ethnocentrism, but . . . only when *both* were present."[56]

The model that Bowles and his colleagues devised assumes that both aggressive and altruistic behaviors are inherited through a single gene, in what geneticists would call simple Mendelian fashion. This is an oversimplification that was necessary to make the complex mathematics of the model more workable. However, Bowles is quick to note that the genetic transmission of the model could be modified to "encompass cultural learning processes and . . . influences of peers and nonparental adults as well as parents."[57] In other words, even if the human capacities for altruism and war are based more in culture than in biology, the takeaway message is the same: Cooperation and conflict evolved together.

Of course, mathematical models are only as good as the assumptions upon which they are based. And it's important to remember that, in terms of being representations of reality, all models are wrong, but some are useful.[58] No one really knows for sure whether the assumptions built into Bowles's model of the coevolution of altruism and war are accurate. No one alive today was there when it happened. However, the intriguing implications of this model are that each tendency was required for the other to evolve. Altruism and cooperation might not have evolved independently without the tendency to become aggressive, competitive, and even warlike. With respect to some of the ways in which evolution has shaped human nature, it seems that opposition was required. In other words, our capacity for good might not have developed without our capacity for evil.

The findings of the Robbers Cave demonstration, the mathematical model of Bowles, as well as the observation of modern hunter-gatherer societies all align with what many know to be an unfortunate truth of human nature. Conflict can serve to unite people within the group. War is the ultimate zero-sum game that fosters out-group hostility and in-group solidarity. Scholars and statesmen alike have understood this for centuries. Indeed, many a shrewd and malicious politician has waged war as a means to score political points and unite the party base.*

Superordinate Goals

After wading through these deep waters, one might be resigned to conclude that war, tribalism, and xenophobia are inevitable. It might be tempting to infer that these are required for in-group unity even today.

* Otto von Bismarck, the Prussian statesman in the late 19th century, understood this principle. Von Bismarck instigated a series of small wars against other European nations (France, Denmark, and Austria) with the aim of unifying the German nation-states into a powerful empire under Prussian leadership.

But, encouragingly, that's not necessarily the case. The final chapter in the Robbers Cave experiment demonstrates this point. Following the brutal between-group competition that developed as a result of the experiment's zero-sum competitive games, the researchers then sought to unify the two groups and erase the sharply drawn lines of group distinction. They tried several approaches. Fostering peace meetings among the leaders didn't work because those who agreed to de-escalate tensions lost status within the group. Sermons on forgiveness and brotherly love seemed to fall on deaf ears. Next, the researchers allowed the groups to interact in noncompetitive venues, hoping this would foster whole-group solidarity. The campers shared meals, watched movies together, and went on intergroup outings. But in each case the boys took advantage of the moment to express their mutual animosity. When they came together for shared activities, seating patterns strictly followed group membership. And the hurling of food and eating utensils in the dining hall was only matched by the hurling of insults.

However, when the researchers introduced a *superordinate goal* that all of the boys could share, intergroup unity started to build. It was necessary that the goal be something requiring cooperation that was important to both groups. To that end, the researchers cut off the water supply to both camps, forcing the Eagles and Rattlers to work together to restore it. For a time, they were all on the same team. Intergroup cooperation was enhanced and conflict reduced. But once the problem was solved, their behavior again degenerated. Successive tasks requiring between-group cooperation were engineered, with further reductions in animosity. Lasting unity was finally achieved as a result of a string of shared superordinate goals: preparing meals together, pooling funds to afford a movie rental, and pulling the supply truck out of the mud. Friendliness between members of groups increased and out-group name-calling was reduced.[59] By this point in the experiment, the Rattlers went so far as to share their prize money from an earlier competition to buy malts for both groups. And at the end of the session, the majority of boys opted to take one bus, not two, back home. The seating pattern, not arranged by the adult leaders, did not follow the group identities of Rattlers vs. Eagles.[60]

The Robbers Cave experiment highlights our simultaneous capacities for aggression and cooperation. Many social scientists have argued that these capacities developed gradually as the overall human population grew, especially in recent centuries, because that growth made competition for resources increasingly necessary. Success in conflict became imperative. Territory became hotly contested. As Richard Wrangham writes, "Cooperation among warriors within groups was so vital for winning conflicts that it evolved to become the basis of humans' exceptional propensity for mutual aid."[61]

Such ideas are based on the assumption that food and other resources are limited, an assumption that originated with Thomas Robert Malthus. A cleric and an economist, Malthus published a profoundly influential work in 1798 called *An Essay on the Principle of Population*.[62] Malthus reasoned that the human population was increasing exponentially while food production was only increasing linearly. Demand for food, he assumed, was going to rapidly outpace supply, which meant that not everyone would survive. Those who did were likely to be more adept at hunting, scavenging, and hoarding food resources. Naturally, this concept heavily influenced Darwin as he formulated his theory of evolution through natural selection. Eventually, the notion that humans had to become aggressive, hostile, and selfish in order to survive came to dominate early evolutionary thinking.

But at some point, things shifted. Malthus turned out to be wrong about many things.*[63] As man domesticated his food sources (such as sheep, goats, and crops), survival no longer became a zero-sum game. In theory, the food sources (which themselves are based on biological populations) could increase at the same rate that human populations did. As farming techniques improved, the main limiting factor for food production was more hands to reap the harvest.** Land was plentiful and natural resources were abundant. In fact, some scholars argue that it is only very recently in our history (just a few thousand years) that there has been an appreciable level of competition between humans. As the anthropologists Jane and Chet Lancaster have argued, "For most of human history, for most of the time, and probably for most people who have ever lived, human experience has been that there is a shortage of group members and that more children and more recruits are desirable for social, economic, and political reasons. Land and critical resources were free and openly accessible."[64] Instead of a nature that is "red in tooth and claw," as Tennyson put it, we might imagine a nature where domesticated plants and animals are abundant and where our ancestors worked together to harvest and share resources.

---

* A big part of the problem was that Malthus's model of population growth did not anticipate advances in technology that have led to more efficient production of goods and services, as well as public health measures that drastically improved life expectancy. Malthus believed the world population had peaked around 1800 and would dramatically fall from that point onward, in part driven by mass starvation. While it's true that hunger and other problems of resource scarcity are global issues today, these are not due to lack of production but rather lack of proper distribution. Yet Malthus's ideas—which in my view are misguided—have resurfaced many times since his days and have influenced, either directly or indirectly, many other economists, policy makers, and opinion leaders. Ironically, in many modern economies, we are now facing the opposite problem, where an aging population cannot be supported by a decreasing fertility rate.

** In more recent time periods, improvements in technology also allowed for new ways to utilize nonrenewable resources.

Going along with this line of thinking, imagine if, at least in some groups poised to compete over precious food resources, a different strategy was chosen? In that scenario, might some groups choose to cooperate rather than wage war? That's what seems to have happened, at least for a portion of our evolutionary history. As culture flourished and hominid intelligence rapidly increased, groups that cooperated could thrive. In contrast, groups that competed against one another were engaging in a zero-sum contest, leaving one winner and one loser.* Those groups that cooperated to achieve superordinate goals fared much better than those who competed against each other. Our species' long history of technological achievements and cultural advancements are evidence that cooperation at some point became at least as imperative as war, if not much more so.[65]

Recent evidence supports the view that superordinate goals can unite people just as much as intergroup conflict.[66] In other words, we tend to cooperate when faced with a task that can only be achieved in groups. A group of humans in a hunter-gatherer society will be motivated to work together to win a skirmish against a rival tribe. However, this motivation stems from the fact that no one person can win the battle alone. The fact that the goal is dependent on conflict may not be that important. The group might be just as motivated to cooperate to construct a new type of housing structure or manufacture a new material. The key ingredient, then, is not that they engage in conflict with another group, but that they engage in an activity where their combined contributions are more than the pure sum of their individual efforts.

It may be true in our evolutionary past that some level of intergroup conflict promoted solidarity within hunter-gatherer groups. However,

* From one vantage point, the struggle for survival is the ultimate superordinate goal. And once domestication of plants and animals was accomplished, survival did not need to be zero-sum.

it is also possible that our ability to cooperate was motivated by the fact that we can achieve more working together than any of us can by working alone.

So which is it? Did intergroup conflict, war, and zero-sum contests lead to our ability to cooperate? Or was it the superordinate goal and the shared purpose (regardless of whether it was zero-sum or not) that allowed human cooperation to flourish? The answer is likely a mix of both. As humans, we are primed to be deeply prejudiced and xenophobic—but we can also be remarkably tolerant of diversity and peaceful even toward strangers. Both capacities are evolutionarily built within us.

The Rewards of Cooperation

One final aspect of cooperation bears mentioning: At some point in our past, cooperation became so important, apparently, that the behavior itself became rewarding even *without* a superordinate goal. Let's return to the work of Tomasello and Warneken, who have conducted social experiments with young children that demonstrate this phenomenon. In one such experiment, an adult stranger engages a young child in a cooperative task that requires the participation of both if it is to be successfully completed. For instance, an object (such as a toy block) needs to be retrieved from a cylinder attached to a table. In order to retrieve the block, the adult needs to lift the cylinder while the child extracts the block. In tasks such as this, as soon as the block has been retrieved, the child will sometimes immediately return the block to its original place and motion to the adult, as if to say, "Let's do it again!" The cooperation itself seems to have been more rewarding than achieving the superordinate goal.[67] At first, this notion of cooperating for cooperation's sake might seem strange. But this is similar to altruism, which, as we saw in Chapter 4, can be rewarding in and of itself.

These findings are intriguing. Apparently, cooperation became so important in our past that the behavior itself became rewarding. We therefore would expect that belonging to some cooperative group would be equally rewarding. Indeed, a desire to belong is a deep part of the human psyche.

A powerful demonstration of this is what happens when humans are excluded from social groups. Most of us can relate to the pain that comes from social exclusion. Interestingly, people feel this kind of pain even when social relationships are new and superficial. By running an online game experiment that he calls Cyberball, the social psychologist Kipling Williams has demonstrated how even virtual strangers can provoke intense emotions. Williams set up the game so that it starts with strangers playing online in groups. Then, after just a few minutes of playing together, one person is excluded. Consistently, he found, the people who get excluded feel anger, sadness, depression, helplessness, and even a reduced sense of meaning in life.[68]

Consider other unpleasant emotions that are attached to our shared sociality. Shame, embarrassment, and guilt—prominent in humans but not as readily observed in animals—all make it possible for humans to show individual commitment to a group when social standing is jeopardized. At least on some level, these emotions can help mend social relationships that have gone awry.[69]

Many scholars, including Christopher Boehm and Richard Wrangham, argue that our moral psychology was forged during a time when becoming a social outcast was extremely dangerous and would almost certainly lead to death. Hence, nature selected *against* the traits that led to social deviance. But it's also very likely that nature also selected *for* traits that enhanced sociality. What if human relationships became so vital that behaviors that led to enduring relationships were also selected for? At some point, altruism, cooperation, love, kindness, empathy, and other prosocial emotions would become extremely important—which is what seems to have happened. We have already seen (in the case of the *Grafton* shipwreck, for example) how some of these traits may have developed because natural selection was operating on a level higher than just the individual.[70]

Regardless of exactly how evolution led to our extraordinary capacity for cooperation, relationships (and the cooperation necessary to sustain them) seem to have been highly prized and valued in human evolution. Humans have evolved such that the harmonious expression of these relationships leads to great satisfaction in almost everybody today. As we will see in Chapter 9, the value and warmth of relationships is the strongest single predictor of personal well-being and happiness. Such psychological architecture is profoundly encouraging.

Chapter Six

Lust and Love

Prairie voles, seahorses, beavers, and bald eagles generally mate for life. Other animals, such as walruses and chickens, are promiscuous. But what about humans? Are we monogamous or promiscuous by nature? Many philosophers and psychologists have concluded that humans are inherently promiscuous and that traditional sexual mores only serve to frustrate and restrict our pleasures and freedoms. Some have even declared that it is against our nature to be monogamous.[1] Is there some truth to this? If so, it's certainly not the whole story.

From an evolutionary standpoint, promiscuity is very easy to explain. An evolving male hominid who impregnated many female partners would obviously reap significant evolutionary benefit. In non-biological terms, our promiscuous behaviors are driven primarily by lust. (A quick note on definitions: I will refer to the propensity to seek numerous sexual partners as promiscuity or short-term mating strategies.)

But what about committed love, or what anthropologists might call long-term pair bonding? This type of attachment, the long-term pair-bonding between men and women, is unusual among mammals. Chimpanzees and bonobos—whom we are told are our closest relatives—have indiscriminate sex. So how did humans come to have the capacity for monogamy? (Another brief note on definitions: for our purposes here, I interchangeably use the terms long-term mating strategies, pair-bonding, monogamy, and attachment to signify this capacity for committed relationships.*)

To be clear, the pattern of long-term commitment and sexual exclusivity among partners does not characterize every human sexual relationship.[2] But the existence of committed relationships in some form is virtually universal among human cultures.[3] In opposition to the capacity for promiscuity, evolution also seems to have produced within us a desire for committed relationships, with an expectation of sexual exclusivity. Our evolved psychological and social capacities reflect this. We have a deep emotional affinity for our spouses beyond just the sexual act. This emotional attachment often causes partners to become infatuated with each other, even to experience a heightened sense of well-being when in each other's presence. In experimental settings, the perception of a painful stimulus is markedly reduced when holding the hand of a romantic partner.[4] In another study, merely looking at a photo of a romantic partner produces a similar analgesic effect.[5]

Other support for our innate tendency to seek monogamy comes from how we respond when a partner betrays a committed relationship through infidelity. Almost reflexively, the betrayed party experiences extreme jealousy, anger, disappointment, or a mix of all of the above. The psychologist David Buss and his colleagues have conducted experiments showing that merely imagining a romantic partner in the arms of another person automatically leads to a powerful response that includes psychological distress, an increase in heart rate, muscular tension, and

* There are doubtless more specific definitions of these terms that psychologists or anthropologists would argue are not the same. For simplicity, I use them interchangeably here.

sweating.[6] The sexual jealousy provoked by infidelity can be extremely powerful and dangerous.*[7] In fact, sexual jealousy is a leading cause of homicide and spousal abuse.[8] Even otherwise normal people can be provoked to rage by the threat of what Buss calls "mate poaching." Summarizing his research that demonstrates the inevitability of sexual jealousy leading to violence, Buss concludes: "In a study of 5,000 people in six cultures, 84 percent of women and 91 percent of men admitted to having had at least one fantasy of murder, and the vast majority fantasized about killing sexual rivals, often in painful and gruesome ways."[9] This near automatic reaction highlights our expectation that there will be sexual exclusivity in a romantic relationship.[10]

At the same time, despite our longings for commitment and fidelity, humans have an opposing nature that is primarily driven by the desire for sexual pleasure and diversity. For biological reasons that we'll delve into, men in particular have the capacity and the desire to be promiscuous in their relationships. In a remarkable way, evolution is responsible for both the drive for long-term commitment and the desire for diversity. Given the jealousies that invariably arise from promiscuous activity, it is difficult to square these two competing desires. On an individual level, each person is tasked with reconciling these opposing impulses. Furthermore, it is impossible for society to satisfy both desires from an institutional perspective. Over time, long-term pair-bonding has proven to be the stronger and more dominant force.[11]

◆

---

* Laws in many cultures reflect the powerful and dangerous response of sexual jealousy, especially on the part of the male. Even up until 1974, in Texas it was legal for a man to kill his wife in the act of adultery. Similar laws have been found throughout many different societies across a variety of cultures.

Robert Sapolsky, an expert in primate behavior at Stanford University, proposes an interesting exercise that illuminates the uniqueness of human mating behavior.[12] Suppose you discover two new primate species: species A and species B. The only thing you know about them is how the males and females differ in their physical appearance. What can this tell you about their mating behaviors? In species A, the males and females are identical in size. In species B, the males are much larger and more muscular than females. In species A, the male and female facial features are fairly similar. In species B, the males have extravagant facial coloration (you surmise the purpose of this is somehow related to mating). With just these few observations we can make fairly accurate predictions about their mating patterns.

Which species demonstrates intense conflict between males for dominance? Species B, where selection pressures have favored larger, more muscular males.

In which species do only a small percentage of the males do virtually all of the mating and reproduction? Species B—this is what all the fighting is about. In Species A, almost all males reproduce a few times each.

Which species has more paternal investment in the rearing of offspring? Species A. Given that almost all males have a few offspring, it's in their best interest to care for and help raise them.[13] In contrast, species B exhibits almost no investment by fathers in the rearing of young.

What is the significance of the extravagant facial coloration of males in species B? It's almost as if to tell the females, "If I can afford to waste all this energy on muscle plus these ridiculous [trappings], I must be in great shape, with the sorts of genes you'd want in your kids!"[14] The females in species B are attracted to these displays of genetic superiority. In contrast, females of species A look for stability, commitment, and a mate who will be a good father.

Which species exhibits monogamous long-term pair-bonding? Species A. Which species exhibits highly promiscuous behavior? Species B.

So where do humans fit in relation to species A and species B? You guessed it. We are somewhere in the middle. Men are about 10 to 20 percent larger than women, with more muscle mass. Men and women have slightly different facial characteristics, with men having larger canines and readily growing facial hair. These examples represent a greater male-female difference than found in species A, but less than the difference of species B. In their behaviors, men clearly have the capacity to be promiscuous, to focus on short-term mating strategies, and to be driven by lustful desires. But they also have the tendency to form long-term, monogamous relationships and to make great investments in their offspring. Evolution has endowed men with both capacities.

◆

Most everyone knows of examples from the lives of friends or family where one marriage partner did not live up to the commitment of fidelity. This is an unfortunate part of human nature. It almost goes without saying that evolution has created within us the capacity for promiscuity. We can easily imagine how the propensity to seek out multiple partners could be advantageous from a Darwinian perspective, especially for males. Presumably, our ancestors lived during some period when a male who impregnated more females had a greater evolutionary advantage than a rival who impregnated fewer. I won't go much further into detail about the promiscuous potential of human nature—enough books have been written on that subject.[15]

But humans are somewhat rare among mammals as creatures who have a propensity for not only promiscuity but also monogamous mating relationships. (Only about 9 percent of mammals form monogamous relationships.[16]) So where did this come from in humans? How did evolution create the capacity for monogamy? No one really knows for sure; no one alive today was there to observe the process as it unfolded. But several theories have been put forward. The theory that makes the most sense goes something like this:[17]

About five million years ago, human ancestors began walking on two limbs instead of four. Becoming bipedal allowed humans to use their hands for things other than locomotion, like constructing and manipulating tools. As this happened, the traits of intelligence and adaptability became ever more important. These attributes allowed humans to use these tools in more and more creative ways to enhance their survival. (I've already highlighted several times that scientists believe that the geologic era immediately preceding our current one was a time of intense and unpredictable catastrophe; intelligence and adaptability proved to be highly beneficial during this volatile era.[18]) These traits required bigger brains. But the problem was that if human brains got any bigger, it would be anatomically impossible for babies to pass through the birth canal of a bipedal woman. What was nature's solution? Keep the brains about the same size at birth, but allow them to continue to develop for an incredibly long time following birth.[*19] This made it possible for the human family to produce remarkably intelligent and adaptable

* The brain development of chimpanzees, macaque monkeys, and other primates slows down dramatically at birth. Humans, on the other hand, have extended brain maturation by a huge amount of time. For example, a six-month-old human has a brain about 50 percent the size of an adult; a two-year-old, about 75 percent; a five-year-old, 90 percent; and a ten-year-old, about 95 percent. In contrast, for chimpanzees, a six-month-old has a brain almost 75 percent the size of an adult; an eighteen-month-old chimpanzee has nearly reached its full adult brain size.

offspring. It also made the human brain incredibly flexible and malleable to the environment.

But this also created another problem: because humans are born with such immature brains, they are utterly helpless and extremely vulnerable when young. They need constant care and attention.*[20] As was the case with other primates, the mother was the default primary caregiver. Being thus occupied caring for her young, the mother couldn't gather enough food or fend off potential predators. She needed help.[21] The most obvious candidate-helper was the father.[22] He had the most at stake evolutionarily in the child's survival. Thus, it became necessary for men to begin investing more time and energy in their children. As this process unfolded and children became more and more helpless and immature at birth, a man's reproductive success could no longer be measured simply by the number of females he impregnated. Instead, it became important for him to see his own children reach sexual maturity, ensuring that he would then become a grandfather. In contrast to earlier epochs of human evolution, it became necessary for men to invest heavily in the rearing of their children.[23] Even today, among modern hunter-gatherers in Paraguay, children without an "investing" father have a significantly higher death rate than children whose fathers are more involved in their rearing.[24] As the psychologist Chris Fraley and his colleagues conclude, "If infants are at greater risk in the absence of the care and protection of both parents, there may have been selection pressures that facilitated pair bonding on the part of mates and a greater degree of parental investment on the father's part."[25] Fathers provided protection and helped teach their children how to hunt, scavenge, and gather. This became an enormous source of evolutionary advantage as

* On average, it takes about thirteen years for a human to reach sexual maturity. It takes bonobos nine years; macaque monkeys, four years; lemurs, two years. The closer related to humans, the longer time period required to reach maturity.

human societies became increasingly complex and survival ever more precarious during the Pleistocene epoch. This invested parenting on the part of the man naturally set the basis for long-term, monogamous pair bonding between the man and the mother of his children. It allowed for humans to develop the capacity for sexual fidelity.

In addition to an increased need for paternal investment in their helpless offspring, other biological changes facilitated this shift toward monogamy. As men invested more of their time and energy in the rearing of their offspring, it became important for a man to know that the children were actually his. If he unknowingly raised a child sired by another man (investing huge amounts of energy and resources into offspring that were not his), this would not have been a very good evolutionary strategy. This is the issue of *paternity certainty*, which can sometimes be a problem for the man. (Due to the way that humans give birth, *maternity certainty* is never a problem for the woman.) Marriage and monogamy—which allowed for repeated sexual union within the pair and restricted sexual interaction outside the partnership—increased the man's confidence in his paternity. This was critical, because at some point during this evolutionary shift, the ancestors of human females concealed their ovulation (the time during which sexual union is most likely to result in pregnancy). In other words, it became practically impossible for a male to tell if his female partner was ovulating. So if she mated with several men in a month, it wasn't at all clear who the father was when the child was born.[*26] This is in contrast to many species of primates, where the female exhibits large red genital swellings during ovulation, which potential partners find attractive. (Approximately 95 percent of all female mammals have some external

* Even in modern times, when men are asked to rate the desirability of over sixty characteristics in a potential long-term partner, faithfulness and sexual fidelity emerge as the traits that were most highly valued.

sign of ovulation, unlike humans.) The concealment of ovulation by human females provided even more evolutionary benefit for marriage and monogamous pair bonding.[27]

Hence, in humans, the capacity for monogamy, engaged fatherhood, enhanced intelligence and adaptability, utterly helpless offspring, and concealed ovulation may have all evolved together. At least that's the best theory to date.[28] Certainly, these traits and characteristics all seem to be related.

So now we have a theory for *why* monogamy developed. But what about *how* it developed? As these changes occurred, many scientists believe that evolution used the capacity for parent-child attachment and adapted it to build the psychological architecture needed for long-term romantic relationships. To review briefly (I discuss this topic in more depth in Chapter 9), parent-child attachment is the critical bond that forms between parents and their young infant. This attachment evolved as humans became increasingly immature at birth and totally dependent on their parents (mother initially, then father later) for nurture, sustenance, and rearing. Because of the helplessness of human infants, this attachment is particularly strong in our species. Basically, everything that an infant does—cry, coo, smile, and laugh—serves to draw parental attention to the baby and strengthen this parent-child attachment. This engenders a powerful response of love, caring, and concern on the part of the parents toward their offspring. If this attachment didn't form, human infants could not survive.[29] As the child grows and begins to crawl and then walk, she explores the natural world, but always using the parent as a secure base. If isolated from the presence of the parent, the child becomes agitated and increasingly distressed, often erupting into fits of crying. Many scholars believe that, over time, this attachment capacity was somehow repurposed to allow for romantic attachment and long-term pair bonding.

What evidence is there that romantic and committed love was adapted from the parent-child attachment? Some scientists hypothesize that because of our prolonged process of maturation, the brain-based system for attachment that is so active in young children never really turns off (in contrast to most mammals). Instead, in humans, the attachment system persists into adulthood, right up to the time when we start to develop romantic interests. Several similarities have been noted between the behaviors that characterize the mother-infant relationship and the bond between romantic partners. For instance, an infant feels safe and more secure when in the presence of their mother. Likewise, romantic partners feel safe and secure in each other's presence. Both mother-infant and romantic pairs often feel anxious or insecure when separated. Both types of relationships are characterized by intimate forms of bodily contact, as well as mutual fascination with each other.[*30] Other evidence supporting this link comes from the fact that the type of relationship you had with your parents in childhood (secure or anxious or avoidant) predicts the type of relationship you will have with your romantic partner.[31] Additional evidence comes from the functioning of our endocrine system.[**32]

Regardless of whether or not our capacity for romantic attachment grew out of our parental attachment, human beings certainly are endowed with the potential to form long-term, committed relationships.

---

* Some scientists even draw similarities between the way mothers speak to their infants in "motherese" or "baby talk" and the way infatuated lovers sometimes communicate (a little bit of a stretch if you ask me, but it's in the literature).

** For example, the hormone oxytocin probably evolved at least a hundred million years ago. In every mammal that has been studied, one of the key purposes of oxytocin is to promote mother-infant bonding. In those rare mammalian species that have developed monogamy (including ourselves), oxytocin promotes long-term female-male bonding. In other words, it appears that oxytocin expanded its portfolio relatively recently (from an evolutionary time scale) to help undergird monogamy.

Not long ago, there was a sense among scholars that romantic love was a cultural product of the Western world. It may be true that throughout human history, a large portion of marriages have been initiated on motivations other than romantic love. It is not true, however, that love is a recent cultural invention specific to Western culture. This myth was dispelled by a landmark anthropological survey of over 160 different cultures from around the world. The authors of this study, William Jankowiak and Edward Fischer, found strong evidence for romantic love with an expectation of commitment in almost 90 percent of these cultures. The data, they conclude, "suggests that romantic love constitutes a human universal."[33]

The concept of romantic love is remarkably consistent across a variety of cultures. Modern global surveys indicate that commitment is central to love, more so than sexual desire. Across both genders, and in most cultures, love is demonstrated by sexual exclusivity (giving up of other romantic partners), sharing of resources, and having children.[34] The vast majority of cultures have formalized this partnership in the institution of marriage, which for most of history has symbolized a joining of sexual, material, and emotional resources into the conception and rearing of the next generation.

◆

Some scholars also believe that female preference among potential mates—what some scientists call female choosiness—may have also played a critical role in the development of monogamous pair-bonding in humans. In other words, over evolutionary time, the female ancestors of modern women developed an attraction toward men who would be more likely to take care of them and their children. This certainly makes sense from an evolutionary perspective, and modern studies provide some support for this idea. In one experimental study, women were shown images of five different men and asked to rank them from most to least attractive. The images consisted of a man standing alone; a man warmly interacting with a young child (smiling at and embracing the child); a man ignoring a young child who is crying; a man and a young child both facing forward and ignoring each other; and a man doing housework. Who do you think was judged the most attractive? The man who was warmly interacting with the child. (The man ignoring the crying child was the least attractive.)[35] Another study with quite remarkable findings: women were able to judge how much interest men (whom they had never met) expressed in children by only looking at a photo of the men's faces. What's more, men's perceived interest in children was rated highly attractive by the women.[36]

In evolutionary terms, this is an example of *sexual selection*. Sexual selection occurs when one sex of a species has a mating preference for certain traits or characteristics in the other sex. This phenomenon can help us understand things in biology that are otherwise puzzling. For example, male peacocks have extravagant, large, and colorful tails that cost an enormous amount of energy to create and maintain. Furthermore, these large tails impede the peacock's ability to flee a potential predator; its ornate colors are also a very conspicuous sign for these predators, almost as if to say, "Hey! Look at me! I'm a tasty peacock who can't move very well because I have this giant tail pinned to my backside!" Why would evolution ever have created such a cumbersome plumage? Well, it turns out, because female peahens seem to prefer large and colorful tails.* And why would female peahens have

* Darwin reportedly expressed that the sight of a peacock's tail made him sick because it didn't fit very well with his original theory of natural selection. His expansion of natural selection to include forms of sexual selection evidently cured his nausea.

this preference? The most likely theory is that it signals a healthy mate. Research has shown that peacocks with large and brightly colored tails have consistently better health than peacocks with smaller tails that are drably colored.[37] Given that health is partly heritable, the offspring of a female attracted to a healthy male would fare better than the offspring of a female attracted to an unhealthy male.

As suggested by the peacock example, sexual selection can apply pressure in opposite directions as other forms of natural selection. Recall from Chapter 3 the case of guppies in Trinidad. In one context, bright colors make guppies more susceptible to predators. Natural selection would exert pressure so that the prevalence of bright colors is reduced in the population because they are not good for survival. But, as it turns out, female guppies are attracted to bright colors and mate preferentially with males who exhibit these traits.[38] Hence, in this context, different forms of natural selection are pushing in opposite directions. Some scholars think that the mate preferences of females (a form of sexual selection) helped contribute to our species' altruistic tendencies (pushing in the opposite direction of selfishness).[39] At the very least, it's likely that the mate preferences of females helped contribute to our species' capacity for monogamous relationships and invested paternal care.[40] Imagine if you were a prehistoric female hominid and had a choice between a potential mate who wanted to add you to his harem or one who showed promise of commitment and would help you raise the children. Whom would you choose?[41]

Other advantages in a shift to monogamy were realized at the group level. To understand how this might be the case, it's important to recognize that the natural ratio of men and women in any given society is one to one.[42] In other words, barring some catastrophe, for every one hundred males in a given society, there will be roughly one hundred females. (This is in contrast to some species, which have an imbalance between the sexes.*[43]) In human groups where monogamy was *not* the norm, even if a small portion of men (say 10 percent) pair up with three to five women each, this would leave a substantial portion of men without potential mates. (In some cases, this could be as much as 50 percent of men without potential mates.) There may have been some circumstances where it was beneficial to a society to have all these unattached, sexually frustrated, and impulsive males, but in most circumstances, an abundance of unmarried men would be decidedly bad for society.** As human social groups shifted to a norm of monogamy, they likely experienced declines in rape, murder, robbery, assault, and abuse. There's also evidence to believe that this shift to monogamy enhanced overall economic productivity, female equality, and paternal investment in children.[44]

◆

As we've seen, over the last three million years or so, human mating patterns shifted from promiscuous, short-term mating strategies to monogamous, long-term approaches.[45] This was likely driven by (or at least associated with) several other changes in our evolution, including enhanced intelligence and adaptability, a prolonged period of maturation for human children, concealed ovulation, more engaged fatherhood, female mate preference, and an overall benefit to society by reductions in violent activity. Of course, as we can all recognize, the shift to monogamy did not eliminate the desire or capacity for sexual

* In one species of plover birds, there are six times as many males as females.

** Another part of sexual selection theory is that members of the same sex (usually the males) will compete against each other—sometimes aggressively—for potential mates. Human societies with an abundance of unpartnered males tend to have high rates of violence, likely because men escalate their competition due to a scarcity of available women.

promiscuity. Humans still have a proclivity toward short-term mating strategies built into our nature. This is evidenced by the desire that many people experience for a variety of sexual partners.

Almost all anthropologists agree, and dozens of studies confirm, that this drive for short-term mating strategies is stronger in men than women.[46] This makes sense evolutionarily. Men have a relatively small biological contribution to the next generation. In contrast, women carry the unborn child for nine months and nurse the baby for years following birth. Because of this, women have a much higher risk of entering into an unfavorable mating union. This makes them much more cautious about casual sex.

Regardless of how it came about, evolution has again left us with competing desires, both of which are impossible to satisfy simultaneously. On the one hand, there is the desire for a variety of partners and sexual diversity. On the other hand, there is the desire for sexual fidelity and exclusivity in a long-term relationship. We are pulled in opposite directions by these very powerful forces, which many authors and poets have described as love versus lust.

Chapter Seven

Free to Choose

It matters not how strait the gate,
How charged with punishments the scroll,
I am the master of my fate:
I am the captain of my soul.
—William Ernest Henley, 1875

Chapters 4, 5, and 6 have demonstrated how evolution has created within us competing natures with respect to several core characteristics: selfishness and altruism, aggression and cooperation, lust and love. To complete my argument and show how an understanding of human nature can reveal the universal purpose of our existence, I must address the concept of free will. The purpose of this chapter is to show that free will is real and that it can be logically and rationally defended from the many attacks currently being made on it. Free will cannot only be readily defended against these arguments, in fact, but in light of everything we have learned so far about the universe and ourselves, the concept of freedom emerges as the most plausible and scientifically sound explanation for a whole host of observations and theories about how humans behave.

Do human beings have free will? This question is as old as history; it has been debated by philosophers, theologians, and scientists for centuries. Freedom, or at least the perception that we are free, is a critical part of human nature. As the writer and neuroscientist Sam Harris puts it: "Most of what is distinctly *human* about our lives seems to depend upon our viewing one another as autonomous persons, capable of free choice."[1] The implications of the debate regarding human freedom are crucial to society and the way we have structured it.[2]

We all certainly feel that we have free will. Life is willful. No one thinks of him or herself as the passive follower of the brain or body. At least to some degree, we each consider ourselves in charge of how we direct and control our attention, thought, speech, and action. This view is shared by people from various cultures of the world.[3] By the age of about four or five, children have developed an intuition resembling that of adults regarding their ability to choose.[4]

Nevertheless, this common-sense view of free will has been challenged by various scientists and philosophers. In this chapter, I endeavor to show that these challenges do not really hold up. In doing so, I hope to show not only that the concept of human freedom can readily withstand these challenges, but also how this traditional view of free will is scientifically sound. Indeed, the notion that humans have free choice is the best way to explain a number of observations from a wide variety of disciplines.

Definition of Free Will

So what do I mean exactly when I refer to free will? The easiest way to answer that question is to unpack the term itself.[*5]

---

* This two-stage definition of free will can be traced back to the iconic psychologist William James.

The first part—*free*—requires that a person acting in the world can make choices about what to do. That person must have *alternative possibilities*. This is also what might be called the "could-have-done-otherwise" principle.[6] For instance, last weekend I chose to play soccer with my son. In order for me to have freely chosen playing soccer, there must have been a real possibility that I *could* have chosen to do something else. In my choice of a career, I could have chosen to be an engineer (which was my undergraduate major) instead of a physician, and as a physician I could have chosen a different medical specialty than the one I did choose. These choices were free because I could have gone down another path. This "could-have-done-otherwise" principle is particularly relevant in the legal system. An assailant is held legally responsible for carrying out an assault because a judge or jury has determined that he not only *could* have done otherwise, but *should* have. The *should* depends upon the *could*.[7] And the law makes that determination because it assumes we have free will.*

The second part of the term—*will*—conveys the idea of *causal mental control*. We can think about what we want to do and then do it; our thoughts and intentions aren't mere byproducts (or noncausal observers) of our actions. Instead, our thoughts and intentions are, at least for some decisions, the root cause.

So, to review: If I have real alternative possibilities and I can control which one I choose, then I have free will. Simple enough, right?

Not everybody agrees.

Nowadays, there are several arguments that scientists or philosophers make against free will. Many of these arguments fall into one of two categories of "-isms": 1) determinism or 2) epiphenomenalism.

---

* In special circumstances, a perpetrator is considered not to have a sufficient understanding or capacity to have done otherwise. In these cases, a person might be found not guilty by reason of insanity. But in most circumstances, the law holds human beings as free and responsible agents.

These roughly correspond to the two requirements for free will to be real: 1) alternative possibilities and 2) causal mental control. I will address both in turn.

Determinism

Determinism is the idea that every event is the inevitable result of previous causes. In a deterministic universe, the present and future are fully determined by the past.* Such a universe would leave no room for human freedom. Everything you do (and are going to do) would be fully determined even before you were born. Whom you marry would be as determined as the date of the next time that Halley's Comet will pass through the inner solar system (which will happen in July of 2061).[8] Determinism is probably the most common (and most loudly expressed) argument against free will today. It corresponds directly to the requirement that in order for free will to exist, there must be alternative possibilities.**[9]

---

* The French mathematician Pierre-Simon Laplace famously hypothesized that if a super intellect could know the position, velocity, and direction of all particles in the universe, then this super intellect could know the future with absolute certainty. This idea is now known as Laplace's Demon, which is based on the assumption that the universe is fully deterministic.

** I should point out here that there are many philosophers (and some scientists) who think we can still have free will in a universe that is completely determined. This line of thinking is called "compatibilism." I must admit that this idea confuses me, no matter how much I read on the subject. If everything in the universe on all levels (including the psychological one) is predetermined by causal events, then I find it hard to make room for anything like free will. Philosophers would classify my view as "libertarian." I tend to agree with the philosopher Shaun Nichols, who says, "It seems like something has to give, either our commitment to free will or our commitment to the idea that every event is completely caused by the preceding events."

As we first address determinism, it's important to recognize that it is by no means a given that the universe is completely deterministic. In fact, there are real examples in nature where it's becoming clear that determinism is *not* the way a given system behaves. The clearest example of this is quantum mechanics, which describes the behavior of objects that are smaller than an atom. For example, if we send a photon (a particle of light) toward a semitransparent mirror, there is a 50 percent chance it will pass through the mirror and a 50 percent chance it will be reflected off the mirror. It seems that nothing in the photon's history affects whether it will pass through or be reflected.[10] In this particular context, the behavior of the photon is not deterministic.[11]

Yet we are also familiar with other systems that seem to be deterministic. Take, for instance, Newton's famous Laws of Motion. If I am driving a car and suddenly slam on the brake, my body keeps moving forward until it is forcefully pulled back by my seatbelt. Similarly, a ball falling in midair does not have a fifty-fifty chance that it will continue to fall. It's going to fall 100 percent of the time. The objects in these examples (the car, my body, and the ball in midair) behave in ways that seem deterministic.

In contrast to complete determinism, our lives seem to have alternative possibilities. As we walk through life, we invariably come to forks in the road. Should I study mathematics or literature? Should I attend this university or another one? Should I buy this house or another? Each choice influences which future we experience. Small choices, such as what to eat for dinner, seem to be of little consequence. However, other choices, such as whom to marry, can significantly affect our life course. The possibility that different life courses are open to us is truly extraordinary. The fact that this is commonly experienced by every living human does not make it any less amazing. And yet, the nonliving objects in our everyday experience all seem to be governed by natural laws, many of which we might describe as deterministic. The planets move in their regular form, acidic compounds combine with basic ones to form water and salt, and autumn always follows summer. Human choices seem an exception to this.

Emergence

To fully address the challenge of determinism, we need to understand the concept of *emergence*, which deals with how different levels of nature relate to one another. Each scientific discipline usually tries to break down a given level into analyzable parts and principles to better understand, predict, or manipulate it. For instance, Newtonian physics deals with the way that objects (specifically those that we can see) interact with one another. Biochemistry deals with the way organic molecules (like carbon, hydrogen, nitrogen, and oxygen) interact with one another to form DNA, proteins, cells, and so forth. Biology deals with how cells form living organisms. Ecology deals with how organisms interact with each other and their environment. Suppose we could see into the nucleus of an atom. Here, we would encounter tiny particles such as quarks, bosons, gluons, and other wondrously named specks of matter. At this level (the subatomic level), the laws of quantum physics are at play. Each discipline deals with a different level and attempts to understand the laws that govern that level.

When we move from one level to another, the laws that govern the first level are not always relevant to the next. A law that governs on the macroscopic level of physics doesn't necessarily apply on the subatomic level.

Sometimes we refer to higher or lower levels. This is usually a reference to the relative size of objects on a given level. For instance, biology, which deals with cells and whole organisms, is a "higher" level than biochemistry, which deals with things like DNA and proteins.

The importance of recognizing that there are different levels to which different natural laws pertain cannot be overstated. In some instances, determinism seems to be a general framework for one level but not another. The chief example of this comes from the two pillars of scientific theory: general relativity* and quantum mechanics.

The theory of relativity seems deterministic. Consider the examples from above on Newton's famous laws of motion (which are a special case of general relativity): most scientists agree that the laws at this level are deterministic. But quantum mechanics is generally considered indeterministic.

These two theories—general relativity and quantum mechanics—are our current best attempts to describe the macroscopic and microscopic world systems. Yet at present, one is thought to be deterministic while the other is not. Part of this seeming incompatibility may come from the fact that these two theories describe physical phenomena at different levels. Hence, if free will seems incompatible with a deterministic view of the world, this could be because human psychology is operating on a level that is not deterministic. The jury is not yet decided on whether a grand unifying physical theory of the universe (whether it is ever found) would be deterministic or not.[12] Perhaps it would have characteristics of both. From our vantage point, it's hard to make much sense of much of the world of nonliving objects without a deterministic framework. But at the same time, it's hard to make much sense of human behavior without viewing each other as free and responsible agents.

Here's where the concept of *emergence* comes into play. Emergence describes how, as we move from a lower level to a higher one, different

* The theory of relativity is a refinement and expansion of Newton's laws of physics and provides a unifying framework for how the macroscopic world behaves.

properties and natural laws are needed to understand the phenomena of the higher-level system. We cannot simply look to the laws of biochemistry to help us describe the principles of ecology. Physicist and Nobel laureate Philip Anderson wrote that "at each [level] entirely new laws, concepts, and generalizations are necessary. . . . Psychology is not applied biology, nor is biology applied chemistry."[13] This is *emergence.* Emergent properties occur when we move to a higher level that takes on properties that are not present in the lower-level system.

Let's take an example from economics. Suppose you could examine all the currency—including dollar bills, quarters, dimes, nickels, and pennies—in the United States. You could weigh the money, examine its physical and chemical structure, and count how many of each denomination are in circulation at a given time. Following this examination, you might be tempted to conclude that inflation is just a myth because you found no evidence of it in your examination.[14] So, too, you would likely not find evidence for the financial concept of interest based on your examination of the physical currency. But you would be wrong. Inflation and interest are real concepts that are important to account for in economics. The concepts of inflation and interest arise when you move from a lower-level system (the physical currency) to a higher-level system (how and why people exchange the currency, which is the field of economics). This is emergence.*

The concept of emergence is extremely powerful but often counterintuitive. Stated another way, emergence means that the whole has properties that are not present in the individual parts. Emergence

---

* With the concept of inflation, though not based in anything physical it is certainly related to physical characteristics. Generally, as more money is printed over time, inflation goes up. But the concept of inflation itself cannot be tied to a given physical object. This is likely the case with free will. Obviously, physical perturbations of the brain (such as infection or intoxication) can affect someone's capacity to choose. But, at least as far as we can tell, this capacity itself is not rooted in any given physical thing in the brain.

describes how "the parts of a system, through their collective activity, can . . . give rise to an entity which is quite distinct—in terms of its structure, its causal powers, its . . . makeup, etc.—from the parts of that system, or from anything those parts compose. The resultant entity . . . is emergent."[15]

Emergence is a concept that is often overlooked or misunderstood in science, even by some scientists. A large part of science involves breaking down phenomena into their parts to better understand the way these phenomena occur. By breaking down things into their parts, we hope to understand the laws that govern their behavior. But the reverse is not always true. Just because we can break down things to their lower-level components does not imply that we can start from the lower level and reconstruct the phenomena at the higher level. If I am striving to create policies to reduce violence in a high-crime neighborhood, the ways in which physicists describe the behavior of a Higgs boson (a subatomic particle) are not really relevant. This is because the laws of particle physics don't have much practical relevance to the principles of sociology. As higher-level systems emerge from lower-level systems, we need new laws, theories, and principles that help us make sense of the way these higher-level systems function.

One of the most remarkable examples of emergence is life itself. Every living thing is essentially made of up carbon, hydrogen, oxygen, nitrogen, and small amounts of other elements. None of these elements have properties that remotely resemble what we know as life. Yet somehow, when combined in the right way, life emerges. In an entertaining exercise, Bill Bryson examines estimates of how much it would cost to buy the raw materials to make a human body. By one estimate, the Royal Society of Chemistry calculates that it would cost over $151,000 for all the carbon, hydrogen, oxygen, nitrogen, sulfur, phosphorus, and other elements needed for a body. By another estimate from a long-running science program on PBS, a measly $168 is all you would need for the raw materials to make a human. But, as Bryson puts it: "No matter what you pay, or how carefully you assemble the materials, you are not going to create a human being. You could call together all the brainiest people who are alive now or have ever lived and endow them with the complete sum of human knowledge, and they could not between them make a single living cell, never mind a [human body]. That is unquestionably the most astounding thing about us—that we are just a collection of inert components, the same stuff you would find in a pile of dirt. . . . The only thing special about the elements that make you is that they make you. That is the miracle of life."[16]

Another one of the most remarkable examples of emergence is how the mind emerges from the brain. The mind, which consists of intangible entities such as thoughts, emotions, feelings, and intentions (i.e., free will) has different properties than the brain, which consists of tangible things such as neurons, synapses, neurotransmitters, and the like. How the mind emerges from the brain is a great mystery. But it does happen, and life as we know it is dependent upon the emergent properties of the mind. It is at these two levels—the lower level of neuroscience (the brain) and the higher level of psychology (the mind)—that our discussion of free will plays out. Clearly, these two levels interact, just as biology and biochemistry interact. Yet the laws and principles that govern each level are not necessarily the same.

◆

A common argument against free will rests on the assumption that our sense of free will is somehow rooted in the brain (which I believe it is). The argument goes like this: the physical brain system is subject to the laws of neuroscience. Because the laws of neuroscience are deterministic, the properties of the mind are also deterministic, thus there is no such thing as free will.

The first flaw in this argument is concluding that an indeterminate level cannot arise from a deterministic one. As the concept of emergence tells us, different levels have different characteristics, including whether the overall framework of that level is deterministic or not. Think about quantum mechanics and Newtonian mechanics. In this case, a system that is indeterminate (at the subatomic level) seems to give rise to a system that is deterministic (at the macroscopic level).[17] The philosopher Christian List has shown that, at least in theory, the reverse can also occur: a deterministic system can give rise to an indeterministic system at a higher level.[18] It is entirely possible that, even if the laws governing the brain are deterministic, the properties of the mind are *not*. As another philosopher argues, commenting on how different levels possess different properties in physics: it is very possible for a given system that it's "micro- and macro-dynamics do not mesh."[19]

But, as it turns out, this complicated (and somewhat counterintuitive) argument may not even be needed. More and more, evidence is accumulating that the laws of neuroscience governing the brain are not deterministic.[20]

Let's consider this a little more in depth. It's long been common to think of the brain as a sort of biological computer. Give it certain inputs and it produces certain outputs. If I show your brain a short video about how a certain brand of potato chips is awesome (the input), the next time you are at the store, you are more likely to buy that brand of potato chips (the output). That's the principle that advertising is based on, and it's why a thirty-second Super Bowl ad costs millions of dollars.

There are many examples from neuroscience and psychology where animals and humans behave in this way: a given input (stimulus) leads to a predictable output (behavior). This makes it tempting to view the brain as a biological computer—as a *passive* system that takes inputs and generates outputs in a wholly deterministic way.

But for more than a century, data has been accumulating that shows that this model is wrong, or at least incomplete.[21] There is something intrinsic in brain systems that generates behavior that cannot be explained by the sum of the inputs. In other words, there is emerging evidence that brain systems (and the behaviors they give rise to) are not deterministic, at least not completely so. This has even been seen in comparatively simple brains, such as those of fruit flies, leeches, and microscopic roundworms.

For example, in one set of experiments, the isolated nervous system of a leech can respond to a stimulus by crawling or swimming. But this can happen despite the fact that the researchers subject the animal to conditions that in every possible respect are exactly the same.[22] In another experiment, scientists will track the flight pattern of fruit flies in response to some external environment. Even when every possible source of variability is accounted for, the fly behaves in a way that, as far as can be observed, is not deterministic.[23] In experiments with the microscopic nematode roundworm (*C. elegans*), researchers will subject the *very same* animal to the *very same* conditions and get different results. While they cannot determine precisely what the animal will do, there seem to be probabilistic laws that govern the animal's behavior. But how a given animal will behave in some contexts appears to be indeterministic.[24]

These findings have given rise to the semi-serious Harvard Law of Animal Behavior, which states that "Under controlled experimental conditions of temperature, time, lighting, feeding, and training, the organism will behave as it damn well pleases."*[25] Martin Heisenberg, the pioneer of neurogenetics, points out that "there is plenty of evidence that an animal's behavior cannot be reduced to responses."[26] Summarizing this phenomenon, one neurobiologist concludes, "Clearly, leeches and flies could and can behave differently in identical environments."[27] The "could-have-done-otherwise" principle noted by philosophers seems to apply even to the central nervous systems of these comparatively simple creatures.** Their past cannot define their future.[28] Would we expect the human nervous system, which is orders of magnitude more complex, to behave in a more simplistic and rigidly determined way? I wouldn't bet on it. Experiments as early as the 1980s have shown that humans can behave in ways that are "indistinguishable from a fundamentally indeterminate process."[29]

To be sure, this does not prove free will. But it does suggest that the level of the brain is not rigidly determined. In our most up-to-date models, the depiction of the brain that neuroscience is creating is one of processes that are permeated with probability, strands of randomness, and indeterminism. Increasingly, the evidence seems to create a framework where alternative options are open to us, even at the level of our brains.

---

* The Harvard Law of Animal Behavior was reportedly developed by students of B. F. Skinner, who, ironically, was an outspoken advocate of behavioral determinism. The students came up with it as they noted that Skinner's behaviorism paradigms couldn't explain all the observations of the animals they used in their experiments.

** The neurobiologist Björn Brembs has started to recognize that some behaviors are deterministic, while others are not. The deterministic behaviors seem to be involved in simple tasks like walking, breathing, and so forth. The indeterministic behaviors are involved in more complex interactions, like exploring/foraging behavior, feeding, or evading a predator.

Causal Mental Control

But it's not enough to show that behavior is indeterministic. Random behavior would not be considered an example of free will any more than rigidly determined behavior would.[30] No educated person would think of the random movements of someone having a seizure as an example of free will. The other important requirement—which corresponds to the "will" part of free will—is to show that our thoughts and intentions are causally related to our behavior. In other words, our mental states (thoughts and intentions) *cause* whatever action we intend to perform. The notion that we have causal control is the mental intention to act, think, or behave in a certain way.

The challenge to causal mental control comes from a belief that is called *epiphenomenalism*. Those who accept epiphenomenalism believe that our thoughts and intentions are merely byproducts of our brains. These thoughts are not causing any behavior, but are simply observers of the predetermined actions of our brains. From the perspective of epiphenomenalism, our mental state is not the cause of anything. Rather it is the molecular machinery of the brain responding to various internal and external factors that leads to a given thought, action, or behavior. My sense of free will is merely a byproduct of non-intentional physical processes that are occurring in my brain. We are a witness to, but not in control of, these physical processes. There is no man upstairs calling the shots. There is no one in the driver's seat. This is the argument against free will from epiphenomenalism.

To be sure, there are of course many things that go on in our bodies that no one would argue are under the conscious control of our mental state. For instance, every second, our bone marrow produces about two million new red blood cells (a truly staggering accomplishment!). Nobody would argue that you cause this process to happen with your mental state. It's a process that your body has (wisely) put on autopilot. There are many similar processes that our body has automated and are thus not under control of our mental state. Digestion, cardiac activity, and maintaining our blood sugar level are just a few examples. None of these processes exhibit causal mental control.

So how do we come to conceptualize mental states as being responsible for those processes—speech, thought, and purposeful movement—that are under conscious control? We intuitively believe that this is the case. No thinking person views him or herself as a mere robot that is at the mercy of the brain. All of us consider ourselves, at least to some degree, in control. But does the science support this view?

The overwhelming answer is yes. In my opinion, the best evidence of this comes from the large psychological literature on goal achievement, mental practice, and simulation. Let's examine just a small part of this literature.

In certain circumstances, imagining oneself performing a certain behavior can increase the actual likelihood that this behavior is carried out in the future. For example, in one study, patients who are arriving at a clinic for psychotherapy were randomly assigned to different groups. In one group, patients were instructed to imagine attending all of the therapy sessions. In the other group, patients were instructed to imagine an irrelevant event (spending time together with a family member) but were also educated about the benefits of attending therapy sessions. The act of imagining oneself attending therapy (even for just a few minutes) cut the probability of prematurely dropping out in half.[31] In this case, the mere act of thinking about an action clearly had some influence on whether that action was carried out.

In many contexts, imagining oneself performing a task with a high degree of accuracy can actually improve performance. For instance, college students were assigned to one of three groups: the first group was instructed to imagine themselves putting a golf ball successfully; the second group was told to imagine themselves putting such that the ball narrowly missed the hole; and the third group was not given any specific instruction. After a week of practice, the group that imagined themselves successful at putting performed the best while the group that imagined themselves missing did the worst.[32] Similar studies have been conducted with similar outcomes in table tennis, dart throwing, football, basketball, soccer, and gymnastics (just to name a few).[33] Mental practice and simulation has also been shown to be beneficial in nonathletic contexts, such as playing a musical instrument,[34] flying a plane,[35] and training for basic medical procedures.[36] A large body of research shows that mental practice and simulation is genuinely effective in improving performance. To be clear, physical practice is also important, and it is likely that in skills that are physical in nature, mental practice has a lesser effect on performance than physical practice. But substantial research demonstrates how the mere act of consciously imagining oneself performing a task well can actually improve one's physical skills. Clearly, conscious thoughts can affect behavior. This is not something we would expect if our thoughts were mere byproducts that had no causal influence on our behaviors.

Similarly, research has shown strongly and consistently that the more detailed a mental plan you have for implementing a goal, the more likely you are to achieve it. This is what psychologists call *implementation intention*. For instance, suppose that I really want to lose weight. I could say to myself (in my mind) that I intend to lose weight. This is a good goal, but it's vague. What the evidence shows is that I will be much more likely to achieve my goal if I tell myself something like this:

> *I will lose weight by exercising before I eat breakfast at seven o'clock every morning for the next sixty days. Furthermore, the type of exercise I will do will be bicycling.*

If I mentally rehearse these details (or commit them to paper), I am much more likely to carry through with my goal (and actually lose weight) than if I merely say to myself, I am going to lose weight. The more I consciously work out the steps toward my goal, the more likely that this behavior will be achieved.

Among a group of college students who were encouraged to exercise, 90 percent of the group that mentally worked out the details of their exercise plan had followed through, while only 40 percent of the group that did not imagine their plan followed through.[37] Two groups of women were asked to perform a breast self-examination in the next month. One group was instructed to imagine the details of their plan to conduct the self-examination; the other group was not given this instruction. All the women in the group that imagined the details of their plan followed through and conducted the exam, while only half of the other group did.[38] In a group of binge drinkers, half were told to imagine ways to limit alcohol consumption (what would they say when offered a drink) while the other half were just told not to drink. The group that mentally planned out the details of their strategy had markedly lower rates of alcohol consumption.[39] In 2006, psychologists Peter Gollwitzer and Paschal Sheeran reported on over 90 independent experiments that show how consciously planning out intended actions has a significant effect on human behavior.[40] If our thoughts were just "bystanders" that did not have any influence over our actions (as the epiphenomenon hypothesis suggests), we would not expect these findings. Clearly, to some degree, we have causal mental influence over what we do.

Mental simulation and practicing are just a few ways in which psychologists have demonstrated causal mental control of behavior. Additional evidence comes from other areas such as mental framing, goal setting, repetitive thinking, and anticipation.[41] Summarizing decades of research in these areas, the psychologist Roy Baumeister and his colleagues conclude: "The evidence for conscious causation of behavior is profound, extensive, adaptive, multifaceted, and empirically strong."[42]

To be clear, unconscious factors can also influence behavior. A number of experiments suggest this is the case. We may not always be fully aware of all the reasons why we do what we do. But just because unconscious motives lead us to do things *some of the time* does not mean that they are *always* steering the ship.[43]

What About Libet?

In the 1980s, a series of experiments led by Benjamin Libet seemingly rocked the free will debate like an earthquake. Even today, many scientists point back to these experiments as disproving our common intuitions. One headline declared in *The Atlantic*, alluding to Libet and similar research: THERE'S NO SUCH THING AS FREE WILL.[44] In light of how much importance was (and continues to be) given to these experiments, I feel some obligation to mention them here.

In the classic Libet experiment, participants were fitted with electrical sensors on their heads that measured their brain waves. They would then be instructed to flick their wrist whenever they wanted to do so. (Specifically, they were told *not* to consciously think about flicking their wrist until they felt the urge to do so.) As soon as they made the decision to flick their wrist, they were instructed to note the moment in time (using a clever, finely calibrated version of a clock) at which they were conscious of the decision to do so. Researchers noticed that participants reported making the decision to move about a quarter second before they actually moved their wrist. However, the electrodes recorded an increase in brain activity (what Libet called "readiness potential") approximately a quarter of a second *even before* people reported being conscious of the decision to move their wrists.[45] This experiment, or some variation on it, has been repeated several times.[46] The conclusion that some scientists have drawn from this is that our thoughts do not have causal power. The brain, they conclude, starts to make up its mind before we are even conscious of making a decision. The sense of free will, they would argue, is just an epiphenomenon of the brain at work.

But let's consider this a little more thoughtfully. There are three events that Libet's experiment attempts to put in chronological order: the conscious awareness of making a choice, the increase in brain activity, and the choice being executed (in this case, flicking one's wrist). A scientifically naïve approach to free will might suggest that these three events occur in the order in which I just listed them: 1) conscious awareness of making a decision, 2) increase in brain activity, and 3) execution of the decision. But a more scientifically informed approach would recognize that the making of a choice involves some form of brain activity. Indeed, it would be very surprising if it did not. From a scientific worldview, free will is based on the physical processes of the brain, even if it cannot be explained entirely in physical terms. Just like thoughts, emotions, and desires, free will is an emergent property of the brain.[47] To expect that the conscious intention to act occurs *before* a change in brain activity would be akin to expecting that emotions could also emerge before a change in brain activity. In a more extreme example, we wouldn't expect that life could emerge before the necessary biochemical compounds (such as DNA, protein, etc.) are arranged into cells, tissues, and organisms. It would be very surprising indeed if this were the case. As the philosopher Eddy Nahmias puts it: "If we assume that conscious processes correlate with [brain] processes, we should expect that conscious experiences do not arise out of nowhere and in no time."[48] The increases in brain activity ("readiness potential") observed in these experiments, it seems, represent the beginning of the decision process, not the end.[49]

There is one final point to make before leaving the Libet-style experiments. In our mental life, a great many choices do not make much difference. Indeed, to simplify our mental life and make room for more important thoughts and decisions, we place many of our behaviors on autopilot. This is true even with some behaviors that are very complex and are sometimes under our conscious control. What I choose for breakfast, whether I step first with my right foot or left, and whether I wear a striped shirt or a solid blue shirt to work are examples of such trivial choices. Most would agree that these decisions do not seem to matter much. Because of the immense complexity of the brain and the physical world with which it interacts, the Libet-style experiments of necessity investigate choices that are inconsequential. We simply do not yet have the tools or the know-how to examine in the same way the more meaningful decisions that humans make. Consequently, Libet-style experiments cannot tell us much, if anything, about the most important forks in the road that all of us face. It is quite a stretch to suggest that the findings of Libet and others concerning trivial decisions (like when to flick your wrist) are applicable to our more meaningful choices (like whom to marry). Let's not forget that participants in such experiments are explicitly instructed *not* to consciously think about the decision beforehand; hence, these decisions cannot be accurate proxies for the meaningful choices into which we put significant and conscious thought, effort, and deliberation.[50] As Eddy Nahmias concludes: "Conscious consideration of alternative options . . . is a wholly different activity from engaging in practiced routines. A body of psychological research shows that conscious, purposeful processing of our thoughts really does make a difference to what we do."[51]

If your mental life is anything like mine, when you contemplate your most important and meaningful decisions, you might not be able to pinpoint the exact moment in time when you made up your mind. For me, these most important and meaningful choices include my decision to marry my wife, our joint decision to have a family, my choice of a profession, and so forth. These decisions were a process more than an instantaneous event. I (and in some instances my wife and I together) wrestled with alternatives, weighed potential benefits, and considered other possibilities. Part of these decisions may well have been unconscious and influenced by my upbringing, culture, and biology. But that doesn't mean that I didn't have a say in the matter.[52]

◆

To review, there are two principal requirements for us to have free will: 1) the existence of alternative possibilities and 2) causal mental control. As we have seen, there is good evidence that human beings meet both of these requirements. In fact, in some form or another, the social and behavioral sciences all presuppose that humans have free will.[53] Far from being merely an epiphenomenon, the notion that humans can causally choose among alternative options is the most credible scientific view to explain a large body of empirical data from social science, to say nothing of the subjective experience of every single human being. In a somewhat counterintuitive way, the concept of free will is extremely powerful in its predictive and explanatory capacity. It would be very difficult to make sense of human behavior without it.[54]

There is one final argument I will make in defense of free will. Some believe that if a group of people are subject to the same harsh conditions, they will all act in the same selfish way. This argument was made by Sigmund Freud. Believing that free will was an illusion, Freud conceptualized that humans were driven primarily by biological urges. If humans were exposed to uniformly harsh conditions, he reasoned, they would all begin to act in a uniform fashion. "Let one attempt to expose a number of the most diverse people uniformly to hunger. With the increase of the imperative urge of hunger all individual differences will blur, and in their stead will appear the uniform expression of the one unstilled urge."[55]

A fellow Austrian psychiatrist, Viktor Frankl, was forced to live the conditions of Freud's hypothetical experiment when he was interned as a prisoner in a Nazi concentration camp. In contrast to Freud's hypothesis, Frankl observed that, even in the terrible conditions of the concentration camp, men could choose to react in different ways. Some of the prisoners, he wrote, started treating their fellow prisoners inhumanely, while others retained a measure of kindness and humanity:

> The experiences of camp life show that man does have a choice of action. There were enough examples, often of a heroic nature, which proved that apathy could be overcome, irritability suppressed. Man can preserve a vestige of . . . freedom . . . even in such terrible conditions. . . . We who lived in concentration camps can remember the men who walked through the huts comforting others, giving away their last piece of bread. They . . . offer sufficient proof that everything can be taken from a man but one thing: the

> last of human freedoms—to choose one's attitude in any given set of circumstances, to choose one's own way. . . . The mental reactions of the inmates of a concentration camp must seem more to us than the mere expression of certain physical and sociological conditions.[56]

Frankl's observations are powerful evidence that humans can exert their free will even in the most difficult of circumstances. Of course, the existence of free will does not mean that our freedom is unconstrained. Anyone will readily agree that we are not free to do whatever we want. We cannot fly. We cannot live forever. There are real biological limits to our freedom. But as we have discussed, these biological influences are often conflicting. In many cases, what we do with these biological influences is up to us:

> To be sure, a human being is a finite thing, and his freedom is restricted. It is not freedom from conditions, but it is freedom to take a stand toward the conditions. . . . Even though conditions such as lack of sleep, insufficient food, and various mental stresses may suggest that the inmates were bound to react in certain ways, in the final analysis it becomes clear that the sort of person the prisoner became was the result of an inner decision, and not the result of camp influences alone. . . . Any man can . . . decide what shall become of him . . . even in a concentration camp.[57]

In the end, I cannot prove beyond a shadow of a doubt that free will is real. This is a debate that some exceedingly brilliant people have been carrying on for centuries. Every reader can believe what they want to believe (you have the free will to do so!). But, as noted in this chapter, there is an immense amount of psychological literature showing that conscious thought affects behavior and that our behavior is not deterministic. In other words, humans have the capacity to choose from among real alternative options, and, at least in many instances, we can use our thoughts and intentions to make those choices. Within a circumscribed set of conditions, we are free to choose. And this idea, as I will argue in the pages ahead, is central to our purpose as human beings.

Chapter Eight

The Meaninglessness of Existence Revisited: An Enlightened View of Human Nature

"A great oddity about humanity is our moral range, from unspeakable viciousness to heartbreaking generosity. From a biological perspective, such diversity presents an unsolved problem. If we evolved to be good, why are we also so vile? Or if we evolved to be wicked, how come we can also be so benign?"

—Richard Wrangham[1]

In 1859, Darwin published *On the Origin of Species*, in which he described the theory of evolution. In this original work, Darwin was mostly silent with respect to humans. Immediately, however, his colleagues began to advocate the view that humans had descended from apes, or that at the very least they shared a common ancestor. In 1871, Darwin published *The Descent of Man*, in which he explicitly applied his evolutionary lens to human origins and behaviors. The idea that humans evolved from lower life-forms was met with—and in many ways continues to cause—great controversy.

Darwin's theory continued to be something of a hypothetical until the turn of the century. In the early 1900s, however, Gregor Mendel's work on heritable traits was rediscovered. And then in the 1950s, DNA was identified as the genetic material, the fundamental unit of inheritance. All of this added tremendous support to evolutionary theory. It was through different mutations and combinations of genes that offspring varied from one another. Those genetic combinations that gave rise to traits conferring a survival advantage became more and more common in the population. This is another way of describing natural selection.

During the 20th century, more and more scientific evidence emerged in support of evolution, but many people continued to resist it as an explanation of the origins and behaviors of humans—at times very publicly (recall the Scopes Monkey Trial of 1925 from Chapter 1). Many still resist it today. Laws against the teaching of evolution in schools have been passed. Some religious groups have firmly opposed and condemned the theory. At every turn, there are people who refuse to see themselves and their behaviors as the products of evolution.

In the 1970s, the Harvard biologist Edward O. Wilson helped pioneer a new branch of biology called sociobiology. Wilson helped launch this new field from its infancy with the publication of his 700-page tome titled *Sociobiology: The New Synthesis*. Wilson and others sought to understand all social behaviors, including human ones, in the context of biological evolution. Several of the important principles discussed in previous chapters—such as kin selection—were developed as this new field emerged. In its broadest sense, sociobiology was an attempt to complete the Darwinian revolution and to unite the biological and social sciences. Wilson tried to do this in *Sociobiology* by carefully detailing the social behaviors of dozens of microorganisms and animals, ranging from ants and termites to snakes and squirrels, and by linking their behaviors with the evolutionary pressures (i.e., natural selection) that formed them. After cataloging and describing over 500 pages worth of animal behavior, Wilson turned his focus to humans. "Let us now consider man in the free spirit of natural history," he wrote, "as though we were zoologists from another planet completing a catalog of social species on Earth. . . . By comparing man with other primate species, it might be possible to identify basic primate traits that lie beneath the surface and help to determine the configuration of man's higher social behavior."[2]

Wilson and other sociobiologists believed that the roots of human behavior were largely biological. As such, they weren't so much developing a new idea as they were synthesizing the existing evidence—hence the subtitle to *Sociobiology*. Wilson and his colleagues were merely cataloging, organizing, and reviewing information that had emerged in the century since Darwin and were using it to argue that evolution, through natural selection, has shaped the behaviors of all creatures, humans included.

Similar to Darwin's *On the Origin of Species*, the ideas of sociobiology generated significant controversy. Backlash came both from within the scientific community and from the lay press. *Time* magazine called the theory of sociobiology "startling . . . and disturbing."[3] At one scientific conference, a group of angry protesters walked up to Wilson and poured a pitcher of ice water on his head, shouting, "You're all wet!"[4] The biologist Stephen Jay Gould, a colleague of Wilson's at Harvard, labeled the ideas of sociobiology dangerous, arguing that similar ideas about biological determinism had been associated with and used to justify several morally repugnant movements, including eugenics (think government-mandated sterilization) and scientific racism (think Nazism). Yet clearly biology does exert a powerful influence on human behavior, and to ignore this can also be dangerous. Mao Zedong justified his radical political regimes based on the view that human nature was a "blank slate."

Sociobiology has undoubtedly helped in important ways to bring together findings and theories from diverse fields such as evolutionary biology, psychology, and anthropology. But there continues to be resistance from the public (and from many social scientists) against a hard biological approach that would attribute all responsibility for our behavior to our genes.* In my view, the largest weakness of sociobiology is that it has not fully incorporated human freedom into its framework of human nature.

Just a few years before the birth of sociobiology, René Dubos wrote a Pulitzer Prize–winning book, *So Human an Animal.* Dubos was a prominent microbiologist who had spent his career at the Rockefeller Institute for Medical Research making important discoveries related to antibiotic development and resistance. In his book, Dubos is highly critical of the way in which biological determinism excludes discussion of free will. "This approach," he writes, ". . . has thrown no light on the nature of human freedom. . . . The ability to choose among ideas and possible courses of action may be the most important of all human attributes."[5] Human freedom, he continues,

> Includes the power to express innate potentialities, the ability to select among different options, and the willingness to accept responsibilities. All these and other such forms of activity involving choice and volition transcend the kind of determinism that would account for the operations of a machine. The very use of the word "machine" in fact

* Of course, as reviewed in Chapter 3, a sophisticated understanding of genetics provides the shocking realization that genes are regulated by the environment!

> points to the conceptual difficulties presented by a narrow view of determinism. . . . Man's body is a machinery of atoms in the sense that its structures and functions obey the laws of inanimate matter. But . . . man himself can control its operations toward certain goals—the life goals that he freely selects. Even among the most orthodox materialists there are few who would not agree . . . that man becomes really human only at the time of decision, when he exercises free will.[6]

Too many biologists have wrongly concluded that evolution robs us of freedom.[7] As Edward O. Wilson himself seems to suggest, if evolution is true, then "we are biological, and our souls cannot fly free."[8] Many such scientists conceive of humans as biologically preprogrammed, with no room for free will.

In a subtle and underappreciated way, this is one of the things about evolutionary theory as applied to humans that has made it unpalatable: We all powerfully experience a sense of free will, so it's not surprising that if we're told that the theory of evolution implies that we don't have free will, then many of us will feel that the theory must not be completely right.

And yet, I would suggest, the idea that free will does not exist because we are biologically preprogrammed is flawed. It does not take into account a great deal of evidence from the study of human behavior. In my view (and as discussed in Chapter 7), the existence of free will is the most logical and scientifically sound conclusion to a very large body of research that has shown that 1) behavior is not rigidly determined (even among relatively simple organisms) and 2) in humans, conscious thoughts and intentions causally influence behavior.[9] Furthermore, as we have seen, if we are biologically wired, we are wired in such a way as to be conflicted. We face an unending series of choices, with biological predispositions that are often in opposition to each other.[10] Free will is a necessary principle for us to be able to tease apart, through the crucible of life, these conflicting biological provocations: selfishness or altruism; aggression or cooperation; lust or love.

◆

Chapter 1 outlined two dilemmas into which an orthodox approach to Darwinism has thrust us: 1) the meaninglessness of existence by virtue of the randomness of our origin and 2) our ultimately selfish natures. Ultra-Darwinists have claimed that our existence has no higher-order purpose other than to continue the struggle of survival and to pass on our genes. We are merely the products of genetic chance and environmental necessity; hence, there can be no higher-order purpose to life. Chapter 2 was my attempt to refute the first dilemma by showing evolution as a process that, in total, was *not* random.

The second dilemma was refuted by Chapters 3 through 6, which show that we are not selfish at our core, at least not exclusively so. Chapter 3 reviewed the different levels on which natural selection has operated in highly social creatures (*multilevel selection*). This is a critical principle because it is one of the key mechanisms allowing for the development of competing natures within us (as well as other social creatures). Due to the way in which evolution has operated—on an individual level and on a kin and/or group level*—we are pulled

* I remind the reader that group selection remains controversial among some scholars. In my opinion, group selection is an elegant and parsimonious way to explain the dual potential of human nature. But even if it is shown that group selection played little or no role in human evolution, we are still left with the observation that humans are wired in conflicting ways. Hence, the whole framework of the book does not stand or fall on the basis of group selection.

in different directions.[11] For our purposes here, I call this the *dual potential of human nature*. Chapters 4 through 6 reviewed in detail some examples of this dual potential of human nature: altruism and selfishness; cooperation and aggression; love and lust. Each side of these opposing characteristics has its root in our evolution. Both cultural and biological factors also have a role in each characteristic.

It should be noted that, in addition to those characteristics explored in Chapters 4 through 6, other examples exist. Upon reflection, it seems that so many of the prosocial aspects of our nature have come about by virtue of kin and/or group selection,[12] while the darker sides of human nature arise from an individual-level selection: greed and generosity; the propensity to lie and the capacity to be honest; the ability to retaliate or to forgive. In a cruel twist of fate, nature has built opposition within us in so many ways.

To complete my argument and infer a higher-order purpose of our existence, Chapter 7 discussed free will. As a basic definition, free will requires an entity (in this context, a person) 1) to have alternative possibilities from which to choose (the "free" part) and 2) to exert some level of control over the decision (the "will" part). Admittedly, I cannot prove beyond dispute that free will is real, but the data reveal that most people intuitively believe that it is. Not only that, but at least based on my reading of the literature, a large and growing base of evidence points to the conclusion that humans have the capacity to consciously choose among alternative (and sometimes opposing) choices.[13] We are free to choose.

In this chapter, we connect the dots and sketch out how the principles discussed earlier in the book point to the purpose of human existence, resolving the existential dilemma of Darwinism. These principles help us link the view of human nature from biology to the view of human nature from the humanities. When considering the *dual potential of human nature* coupled with the principle of free will, the higher-order purpose of our existence comes into focus with remarkable originality and yet haunting familiarity. This purpose is to choose between these competing natures: to choose between selfishness and altruism, between aggression and cooperation, between love and hate, between promiscuity and committed pair-bonding, between forgiveness and retaliation. In the language of an ethos that has largely been discarded, a fundamental purpose of our existence is to choose between the good and evil natures inherent within each of us.

Throughout history, many humanists, philosophers, and theologians have recognized this *dual potential of human nature*, often using different language. Giovanni Pico della Mirandola, the Italian philosopher and humanist of the Renaissance, wrote the *Oration on the Dignity of Man*, which expressed the spirit of humanism at the dawn of the Renaissance. In this classic discourse, he described human nature in this way:

> The Great Artisan created man . . . and then told him: you . . . shall determine for yourself your own nature, in accordance with your own free will. . . . You shall have the power to degenerate into the lower forms of life, which are brutish. But you shall also have the power, out of your soul's judgment, to be reborn into the higher forms, which are divine.[14]

Shortly after Darwin's theory of evolution became public, the Scottish author Robert Louis Stevenson wrote a famous novella, *The Strange Case of Dr. Jekyll and Mr. Hyde*. One of the central themes of the story is the idea that every person has the potential for good and evil within him or herself. Dr. Jekyll is a well-respected, intelligent scientist who has "every mark of capacity and kindness."[15] However, Jekyll has a dark personality against which he wrestles, Mr. Hyde, who commits violent acts against the innocent. Mr. Hyde's characteristics are those of selfishness and aggression.[16] The book was a huge commercial success, selling 40,000 copies in the first six months. Perhaps one of the reasons for the novella's success is that it epitomizes the struggle that all of us experience.[17]

The Extremes of Human Nature

In addition to providing for purpose, this enlightened view of human nature also helps us escape the quagmire that philosophers, economists, and scientists have been stuck in for centuries. Are we selfish or altruistic? Are humans aggressive or prone to cooperate? Are we monogamous or promiscuous? The stunning answer seems to be . . . yes. Or at least we have the potential to be all of these things. This helps explain the huge variation in human nature—our great "moral range"—that has been chronicled throughout history. This view of human nature is also consistent with the observations of scholars who have argued that there is no purpose to our existence. For example, the writer and neuroscientist Sam Harris, who is a vocal atheist, has written, "While no other species can match us for altruism, none can match us for sadistic cruelty either."[18] This paradox is a consequence of true freedom coupled with the way in which evolution has created competing predispositions within us.

The annals of history are sullied with examples of the darker propensities of human nature. One hardly has to try to come up with cases in point of this unfortunate truth. Murderers, rapists, thieves, despots, tyrants, and dictators have wreaked havoc and inflicted unspeakable suffering on their fellow humans. Their existence proves the unfortunate reality of the immense capacity for humans to behave in terrible ways.

Yet history is full of moral heroes, too. Those who have chosen and heeded the quiet but stirring and unselfish appeals to human nature. These examples are a reason for hope and, I believe, a reason that this view of the dual potential of human nature is so illuminating.

Consider Barbara Goodson, for example, who in 2015 started giving free haircuts to the homeless. At first, just a few cuts per month. Then she started giving haircuts to people coming out of prison. Next, she included battered women. This soon turned into hundreds of haircuts. Eventually, she helped form a nonprofit organization that continues to grow. What evolutionary pressure or tendency led her to render service to the outcasts of society? To those who could likely never repay her and were not blood relatives? She chose to heed the unselfish part of her nature, derived from several evolutionary mechanisms, but she chose to apply it broadly, even to people who might never be able to reciprocate.[19]

As a medical student in Baltimore, I witnessed firsthand the altruism of a remarkable and energetic young woman named Sarah Hemminger. Sarah started what was then called the Incentive Mentoring Program (now called Thread). This program connected volunteers (the majority of whom were graduate students) with high school students from the local school district at risk of dropping out.[20] Over a period of more than fifteen years, this initiative has helped hundreds if not thousands of high school students graduate and receive higher education. What was it that led Sarah to undertake such a tremendous effort? Again, she (like all of us) has the capacity for such altruism within her. But she (unlike many of us) simply chose to express this altruism in an incredible way.

Sometimes the inspiring power of the goodness of people comes in response to tragedy. The capacity to forgive has its roots in the importance that evolution has placed on relationships. It allows for the mending of damaged relationships to permit remorseful wrongdoers to be admitted back into the social circle. Yet this capacity is often opposed by intense feelings of moralistic aggression and the desire to see the perpetrator punished.[21]

Charles Carl Roberts was a milk truck driver who delivered dairy products to the Amish communities in rural Pennsylvania. In October of 2006, for reasons not fully known, he seems to have lost all control and reason. He forced his way into an Amish school, dismissed the boys and the adults, and tied up ten girls. He shot the girls, killing five and wounding the other five. He then shot and killed himself. In the face of this unspeakable tragedy, the Amish community responded in a most remarkable way: they promptly chose to forgive. That very evening, the families of the victims sent words of comfort to Marie Roberts, the killer's widow, who was undoubtedly stricken with grief, embarrassment, guilt, and confusion. The Amish set up charitable funds for Marie and her three children. In an open letter to the Amish community, Marie later wrote, "Your love for our family has helped to provide the healing we so desperately need. Gifts you've given have touched our hearts in a way no words can describe. Your compassion has reached beyond our family, beyond our community, and is changing our world, and for this we sincerely thank you."[22]

In the fall of 2004, forty-four-year-old Victoria Ruvolo of Long Island, New York, was driving home from seeing a niece sing at a recital when the eighteen-year-old passenger of an approaching car, Ryan Cushing, hurled a twenty-pound frozen turkey toward her oncoming vehicle. Cushing intended the act as a prank, but the frozen turkey shattered the windshield of Ms. Ruvolo's car and fractured many of the bones in her face (including her left eye socket). After extensive reconstructive surgery, Ms. Ruvolo spent nine months in the hospital and rehabilitation centers before returning to work. Suffolk County prosecutors were seeking the maximum sentence of twenty-five years in prison for Mr. Cushing. But Ms. Ruvolo had other plans. "I didn't want Ryan to rot in jail," she later recounted. She lobbied for a more lenient sentence, hoping that he could still make something of his life. After pleading guilty, Cushing stopped to speak with his victim in the courtroom. The two embraced as he wept. "I'm so sorry. I didn't mean it." Her reply: "It's okay. I just want you to make your life the best it can be." Two months later, when he was sentenced to six months of jail and five years of probation, he addressed her in the courtroom: "Your ability to forgive has had a profound effect on me. It has already made a positive change in my life." As of 2019, when Ms. Ruvolo passed away, Mr. Cushing was employed and a productive member of society. Through involvement in a nonprofit organization, he had spent significant time speaking to youth about overcoming their mistakes and avoiding them in the first place.[23]

Like many crises, the COVID-19 pandemic brought out the very best but also the very worst in human nature. At the time the pandemic hit the United States, Dr. Richard Levitan was practicing emergency medicine in rural New Hampshire. With the skills and capabilities to help manage patients who lost the ability to breathe on their own, he had an unselfish desire to help where he was most needed. He heeded the plea of the governor of New York, whose state was hardest hit during the beginning phases of the pandemic, to come offer his clinical skills in New York City. Dr. Levitan had a brother who lived in the city, and, after trying unsuccessfully to find temporary residence in hotels, he decided to stay at his brother's place. When word in the apartment building got around of why his brother was there, the building's board of directors decided they did not want him there. Nearing the end of his first shift in the emergency room, his brother had to text him:

"Hey Richard—We are so proud of you and your heroism. I hate to be the bearer of bad news but looks like our apartment building doesn't want you staying in our [apartment]." As the *New York Times* journalist who chronicled Dr. Levitan's dilemma concludes, crisis moments "can make ordinary people turn valorous or villainous."[24] Upon reading this story we are simultaneously touched by the altruism of Dr. Levitan and enraged by the selfishness of the unnamed individuals who formed the apartment's board of directors.

How do we explain the behaviors of the heroes in these stories? In each case, they faced a choice of how to respond, and each could have chosen otherwise. To act unselfishly, to forgive, to treat others with compassion—these capacities are nested to some degree in all of us. But so are the capacities to act with selfishness, aggression, and disregard for the feelings of others. Clearly there is variation in these capacities between different individuals. And clearly, certain cultures and contexts can appeal to one of these natures more than the other. But in a most remarkable way, nature has built within us competing and opposing characteristics and somehow endowed us with the ability to choose between them. How we have been given the freedom to choose is still a great mystery,[25] but in my view it is clear that humans have this capacity to some degree.[26]

Most of the decisions we face do not come in the midst of such terrible personal or national tragedies. But their cumulative effects are nonetheless important and shape society (as well as our life outcomes) in profound ways. Moving beyond these heroic examples of altruism, forgiveness, and compassion, there are "innumerable smaller impulses of courage and generosity that bind societies together."[27] Our lives and our communities are shaped by thousands of small yet significant decisions, the most important of which are centered on whether humans behave selfishly or altruistically.

Integrating the human capacity for freedom with the principles of the dual potential of human nature provides for a more enlightened view of human nature. It also provides a framework for a higher-order purpose for our existence: Given that evolution has fashioned within us opposing characteristics and that we are ultimately possessed of free will, *a principal purpose of our existence, it would seem, is to develop our ability to choose between our competing natures.* Life seems to be a test. Our purpose, at least one of them, is to choose between the good and evil within us.

How Free Are We?

Now, some scholars might reject this framework of human nature and purpose because of evidence suggesting that some of our behavior is not under our conscious control. They might argue that we can't choose to be altruistic any more than we can choose to fly. It is true that some of our behavior might best be characterized as automatic, intuitive, emotional, or subconscious, and therefore not directly under our conscious control.[28] But many behaviors *are* directly under our conscious control. The trick is to place ourselves in contexts that elicit our better nature.

The psychologist Jonathan Haidt has developed the useful metaphor of a rider on the back of an elephant. The elephant represents our automatic, intuitive, or emotional behaviors that we cannot (easily) control. The rider represents the rational and deliberate parts of our behavior that we can control. As Haidt describes: "I'm holding the reins in my hands, and by pulling one way or the other I can tell the elephant to turn, to stop, or to go. I can direct things, but only when the elephant doesn't have desires of his own. When the elephant really wants to do something, I'm no match for him."[29]

Haidt gives the example of avoiding dessert. When dining at a restaurant, Haidt reports having ultimate authority over controlling whether or not he orders a given dessert from the menu. In this way, he is controlling the elephant. But if the dessert is placed on the table in front of him, the elephant takes over and he is seemingly powerless to resist. Most of us can relate to experiences such as this in which we really wanted to modify our behavior (and may have made several attempts to do so) but failed.

Psychologists may be right when they assert that our ability to control ourselves is perhaps not as great as we intuitively believe. In other words, our free will is perhaps not as free as we think. But we do have some control. The trick is to lead our elephant to places where he will be influenced to be on his best behavior. At the very least, we should avoid leading him to places that bring out the very worst in our natures.

The choice of whether or not to eat dessert can be significant. Over time, repeated indulgences in dessert can lead to poor health. But there are many other examples that are even more important to our personal lives and to society. For example, if you are trying to kick a drinking habit, it's best not to let your elephant wander into a bar. Once inside the bar, it's going to be *very* difficult to control an alcoholic elephant. The best course of action is to prevent the confrontation with temptation from the outset. An ounce of prevention is worth a pound of cure.

How about violence? What kinds of contexts elicit violent responses and how can we prevent the elephant from wandering into these places? If you recall from Chapter 6, sexual jealousy is a leading cause of violence (and of one-on-one homicide). From an evolutionary perspective, this makes sense. Given the way we have evolved and how much we have to invest in our helpless children, it is almost inevitable that the reaction to learning of a partner's infidelity is one of extreme anger, jealousy, sadness, and disappointment.

Drawing on this body of research and continuing Haidt's metaphor, if we find ourselves riding an elephant that has wandered into a room that elicits sexual jealousy, it will be very difficult to control. The elephant will likely experience a powerful emotional response, which can lead to intense psychological distress in the betrayed and can result in deadly consequences for the betrayer.* Marriage, it would seem, is one of the solutions to keeping this part of our nature in check. In the analogy of the elephant and rider, marriage is one of those places that will help influence the elephant to be on his best behavior. Later in the book, I'll go into more detail about how marriage can influence behavior for the better in other important ways.

◆

This chapter brings together the main principles from Chapters 2 through 7 to resolve two Darwinian dilemmas. Our coming into existence was not random (Chapter 2), allowing for the possibility of an overarching purpose to our lives. But evolution seems to have shaped us in such a way that we are pulled in different directions: the dual potential of human nature (Chapters 3–6). We also find ourselves in possession of free will (Chapter 7). As previously noted, when we bring it all together, it seems clear—at least from my vantage point—that life is a test. A principal purpose of our existence is to choose between the good and evil inherent within us.

---

* This seems to be one of the principal reasons that God has placed strong moral sanctions on the use of sex. All religions teach some form of sexual restraint.

Given that our natures are strongly influenced by both biology and culture, it is reasonable to conclude that some contexts are better than others in nudging us to choose the better angels of our nature. In the subsequent chapters, we explore what I believe are the most important social arrangements that help us to choose our better natures at the same time as maximizing individual well-being and happiness.

Chapter Nine
The Good Life

In 1937, Arlie Bock, then the director of the Harvard University Health Services, met with the philanthropist William T. Grant and proposed a study on health and well-being. Bock concluded that too much medical research was devoted to the study of disease and not enough to studying wellness. In other words, more research was needed on what made people healthy and happy rather than what made them sick.[1] One year later, Bock began what became known as the Harvard Study of Adult Development—one of the longest-running studies into human health and well-being. The original question the study proposed to investigate was this: What factors lead to a good life?

The study's research team consisted of psychiatrists, psychologists, internists, physiologists, and anthropologists, and its initial group of volunteers—numbering 268—was drawn from among the healthy young Harvard sophomores.* All of the volunteers were male, because Harvard was not yet enrolling women. In time, Bock's study would

---

* Many notable men emerged from this study cohort, including a future president of the United States (John F. Kennedy), a man who served in a presidential cabinet, and the long-time editor of the *Washington Post.* (At age 74, Ben Bradlee self-disclosed his participation in the study with the publication of his memoir, *A Good Life.*)

join with another study, begun a year later, that followed a group of teenagers—nondelinquent adolescents from the poorest neighborhoods of Boston—over their lives. Altogether, more than 700 men were followed by the study's researchers for over seventy years.[2]

These men agreed to repeated and rigorous evaluations. In the initial assessment alone, they spent twenty hours in testing and evaluation. This included eight hours with a psychiatrist, a thorough interview with a sociologist (who also went to various parts of the country to interview their families), a two-hour physical exam, an extensive profile of physiologic characteristics (including the 1930s version of a stress test, measures of respiratory function, and an insulin tolerance test), and the Scholastic Aptitude Test, or SAT. They interpreted Rorschach inkblots, submitted handwriting samples for analysis, and underwent electroencephalogram evaluations (a new technology at the time, which the researchers hoped would reveal correlations between brain wave patterns and personality).[3]

From the 1970s until 2005, the psychiatrist George Vaillant led the study, and during his time as its director he concluded that the evidence was clear: much of the mental health and well-being of the men was derived from their relationships with family and friends. When one wealthy and successful participant in the study was asked to what he attributed his vitality, he replied, without hesitation, "My wife and family." This had surprised Vaillant, who had initially assumed that the man's well-being stemmed from his financial achievements. "I saw him as a smashing business success without emotional limitations," Vaillant said. "But [to him], his *family* were his jewels."[4]

Over time, Vaillant and others recognized that the ability of the men to form and sustain enduring relationships predicted very much about mental and physical health, happiness, and life satisfaction. Those who had difficulty forming or maintaining relationships were much more likely to seek psychiatric or medical help and were more likely to develop chronic physical illness by middle age.[5] Overall, a strong connection was found between the warmth of the men's relationships with their families and how well they fared in many other aspects of their lives.

When asked at the conclusion of his career what he learned from the study, Vaillant offered this take: "The only thing that really matters in life are your relationships to other people."[6] Vaillant's successor, the Harvard psychiatrist Robert Waldinger, has summed things up in a very similar manner: "The clearest message that we get from this 75-year study is this: Good relationships keep us happier and healthier. Period."[7]

Attachment Theory

By modern standards, the Harvard Adult Development Study has a number of weaknesses. Its analytic methods were somewhat unsophisticated. Until relatively recently, it did not seek to include the perspectives of women. Its definitions of happiness and satisfaction (concepts that psychologists have only recently attempted to define with scientific precision) were somewhat vague. However, its strength lies in the extremely long-term follow-up of its study subjects. In this respect, it is matched by few other research endeavors—and, significantly, a large body of research has confirmed its central finding: that relationships are critical to human happiness.

Some of the most important aspects of that additional research was conducted by the British psychiatrist John Bowlby, who devoted his career to the study of the importance of human relationships. When Bowlby was just starting his career in the 1930s, Sigmund Freud's theories of psychoanalysis still held great influence. Freud had insisted that relationships were merely a means to an end: they made possible the satiation of physical drives. The infant loved the mother because she gave milk. The husband loved the wife because she gave him sex. In this school of thought, there was little recognition that humans could have social needs beyond the means to satisfy physical ones. But Bowlby felt that didn't do justice to the full depth and complexity of human social interactions.[8]

Searching outside his own field of expertise for evidence of the importance of relationships, Bowlby met and befriended several experts in animal behavior.[9] These experts, he came to realize, never assumed that physical needs were the primary drivers of animal behavior.[10] "On the contrary," he wrote, "all their work has been based on the hypothesis that in animals there are many in-built responses which are . . . independent of physiological needs and responses, the function of which is to promote social interaction."[11]

Bowlby also found evidence in the work of a fellow psychiatrist, René Spitz, who made important observations on the outcomes of orphaned children.[12] In the 1940s, when Spitz was conducting his research, orphanages were working hard to reduce an abysmally high mortality rate. Because so many of the deaths were due to infection (antibiotics had not yet been developed), many orphanages went to great lengths to ensure that their facilities were clean. Unfortunately, in their efforts to create a sterile environment, they went too far. Extreme measures were taken to ensure that children had little to no contact with anyone or anything that could be a carrier of germs.

The orphanages were also likely influenced by the erroneous views of the era's child-rearing experts, many of whom warned that too much mothering could be detrimental to proper early development. John B. Watson, a very influential psychologist who served as president of the American Psychological Association, urged mothers to adopt this approach to their children:

> Never hug and kiss them, never let them sit in your lap. If you must, kiss them once on the forehead when they say goodnight. Shake hands with them in the morning. . . . Won't you then remember when you are tempted to pet your child that mother love is a dangerous instrument? An instrument which may inflict a never-healing wound, a wound which may make infancy unhappy, adolescence a nightmare, an instrument which may wreck your adult son or daughter's vocational future and their chances for marital happiness.*[13]

(It's sometimes astounding to think that views like these were ever seriously considered legitimate.)

In studying orphanages that had adopted these approaches, René Spitz observed that, almost without exception, the children institutionalized there developed serious social and psychiatric problems. And the reason, he concluded, was obvious: by so dramatically limiting the children's social interactions, the orphanages had completely "sterilized" their psyches.[14]

To confirm this hypothesis, Spitz conducted a seminal study in which he observed two groups of infants during the first years of their lives. The children in the first group were raised in a traditional (sterile) orphanage, and those in the second group were born to "delinquent minors," young mothers who were incarcerated at the time their babies

* After one of Watson's lectures, a "dear old lady" approached him and said, "Thank God, my children are grown—and that I had a chance to enjoy them before I met you."

were born. These babies remained with their mothers in a prison environment for the first year of life.

In some respects, these groups of infants lived in similar environments. The food, clothing, suburban location, physical layout, and hygienic conditions at each institution were alike; if anything, some of the conditions in the orphanage were superior. The key difference between the groups of babies was their access to social interaction. The prison mothers, who were deprived of most activities, spent almost all of their time doting upon their new babies. On the other hand, the infants in the orphanage lacked human contact most of the day. After one year, the differences between the two groups were striking. The prison infants had flourished, while the infants in the orphanage had languished and were exhibiting substantial delay in cognitive and developmental milestones. In all respects—physical, intellectual, social—the infants of imprisoned mothers fared much better than the children of the orphanage. As Spitz concluded: "We believe they suffer because their perceptual world is emptied of human partners. . . . The result . . . is a complete restriction of [mental] capacity by the end of the first year."[15] Thirty-seven percent of the children from the orphanage died. The vast majority of the rest suffered severe developmental delay. In contrast, no child from the group of incarcerated mothers died.[*16]

By the 1940s and 1950s, some psychoanalysts were beginning to recognize that social relationships were more important than they had previously assumed. Drawing upon these insights, Bowlby brought together observations from Spitz and from experts in animal behavior to develop a theory that humans have deep social needs, even

* Spitz concluded that "the damage inflicted on the infants in [the orphanage] by their being deprived of maternal care, maternal stimulation, and maternal love, as well as by their being completely isolated, is irreparable."

independent of biological ones. In 1958, he brought all this together in a seminal paper titled, "The Nature of the Child's Tie to His Mother." This was the beginning of his theory of attachment, which he fully expanded in the following decade with the publication of his now famous three-part text, *Attachment*.

Bowlby's theory of attachment incorporated insights from evolution. Human newborns, he observed, cannot do anything for themselves. Everything a baby can do—crying, cooing, sucking, clinging, smiling[17]—is done to enhance its relationship to its parents (especially its mother) and keep them close by. This is evolutionarily adaptive, Bowlby reasoned. Without the parental bonds fostered by these primitive behaviors, the baby would die.*

But the behaviors of the baby are just one side of the coin. In order for the baby to survive and even thrive, it must foster a strong reciprocal attachment on the part of the parents. Which it does. Anyone who has had children can remember what profound love is experienced upon seeing your child's first smile, laugh, or coo. The baby, which represents a huge evolutionary investment for the parents, engenders a strong sense of love from necessity.[18]

The journalist and social activist Dorothy Day described this experience as follows: "If I had written the greatest book, composed the greatest symphony, painted the most beautiful painting or carved the most exquisite figure, I could not have felt the more exalted creator than I did when they placed my child in my arms. . . . No human creature could receive or contain so vast a flood of love and joy as I felt after the birth of my first child."[19] Many parents can relate to this. The birth of a first child can induce within a parent a depth of joy and

* Though he didn't necessarily label it as such, this is an aspect of kin selection in humans.

love that was previously not fathomable. It's like you discover another chamber in your heart that you didn't know existed. This is because of the way in which evolution has shaped us. Family relationships represent the most powerful forms of love and affection that nature has created.

As mentioned in Chapter 6, compared to other mammals, human infants have an extremely long maturation process; they are born utterly helpless. Some experts in child development even refer to the first three months of life as the "fourth trimester."[20] Indeed, compared to other animals, human brain growth progresses more slowly following birth. This means that human infants are extremely immature for many years following birth.[21] Giraffes can walk within an hour of being born. Foals can gallop when only a day old. But human babies are utterly helpless and depend upon their parents for protection and provisions for many years.

But the slow rate of human brain development has its advantages. It allows us to be shaped by our postnatal environment more than any other creature. Nature seems to have struck a delicate balance. To be able to walk upright, human pelvises narrowed over the long course of evolution. Forty weeks of pregnancy produce an infant that is about as large as possible given the necessary passage through a woman's birth canal.[22] Even though infants may not appear to be very social creatures shortly after birth, extraordinary changes in the brain are happening during this period that lay the foundation for almost all subsequent social interaction.[23]

Since Bowlby's time, a growing body of research supports this notion that the mother-infant attachment lays the groundwork for all other relationships. Bowlby himself considered the behaviors that nurture strong relationships in later life as a natural extension of the behaviors strengthening the attachment between mother and child. "Attachment behavior in adult life," he wrote, "is a straightforward continuation of attachment behavior in childhood. . . . In sickness and calamity, adults often become demanding of others; in conditions of sudden danger or disaster a person will almost certainly seek proximity to another known and trusted person. In such circumstances an increase of attachment behavior is recognized by all as natural."[24]

As discussed in Chapter 6, many scholars also believe that the psychological architecture that makes possible the mother-infant attachment was repurposed to allow for long-term, monogamous romantic relationships.

Positive Psychology

Today, it's well established that children have a deep need for parental love and attachment. But this proves only that the lack of warm relationships is detrimental to the well-being of children. Do relationships also sustain our happiness and well-being into adulthood? Does the central finding of the Harvard Study of Adult Development hold, notwithstanding the study's weaknesses? What does modern psychology tell us about relationships, happiness, and well-being?

One place to look is the field of positive psychology, which aims to understand what enables well-being. One of the field's founders is Martin Seligman, a professor of psychology at the University of Pennsylvania. According to Seligman, well-being (a term he prefers to happiness) is made up of the following key elements:

- *Positive emotion*. This concerns just what you think it does: pleasant, happy, and satisfying emotions.
- *Engagement*. This involves being so engrossed in doing something that you lose track of time. Psychologists

sometimes call this the state of "flow." All of a sudden, you look up from whatever you are doing and several hours have passed without you noticing.

- *Meaning.* According to Seligman, this involves "belonging to and serving something that you believe is bigger than the self."[25] Meaning is almost always tied to how we interact with others.[26]
- *Accomplishment.* This often involves doing things just for the sake of doing them. It can satisfy important needs, but taken to extremes it can also lead to a deterioration of personal relationships.
- *Positive Relationships.* These, again, are just what you think they are: warm and nurturing relationships, especially with close friends and family, which give us a sense of peace, happiness, security, and well-being.

In theory, these five elements are defined and measured independently, but in reality, all of them tie back to relationships. "Very little that is positive is solitary," Seligman writes. "When was the last time you laughed uproariously? The last time you felt indescribable joy? The last time you sensed profound meaning and purpose? The last time you felt enormously proud of an accomplishment? Even without knowing the particulars of these high points in your life, I know their form: all of them took place around other people. *Other people* are the best antidote to the downs of life and the single most reliable up."[27]

But there is a catch. Just as sure as good relationships lead to happiness, bad relationships make people miserable. Most people have experienced this in some form. Ongoing conflict with a colleague, a roommate, or (worst of all) a spouse can make life terrible. Think of it this way: Although it's true that there's nothing better for your happiness than a good marriage, there's also nothing worse than a bad one.[28]

Marriage and Well-being

Marriage represents a unique and special form of human relationship. When two people come together in marriage, in virtually all cultures for most of history, it has symbolized the commitment of emotional, material, and sexual resources to the marriage partnership and to the rearing of the next generation. How does marriage affect well-being and happiness? In the last several decades, some have portrayed marriage as restrictive and limiting, sometimes implying that marriage might even reduce freedom and happiness. But what does the research actually say?

In their pivotal report from 1976, *The Quality of American Life*, the scholars Angus Campbell, Philip Converse, and Willard Rodgers found that the most important areas that lead to life satisfaction and happiness were a good marriage and family life.[29] Since this time, dozens of studies have confirmed that marriage plays a central and positive role in happiness and well-being. One report, drawing on six years of data from a national survey, found that marital satisfaction played a "far greater" role in a person's overall happiness than any other endeavor or activity, including work, friendship, health, and financial success.[30] Another study of over 2,200 adults in the United States found that married men and women reported more happiness and less depression than their counterparts who were unmarried.[31] In another analysis, marital happiness was the most powerful predictor of general life satisfaction and mental health.[32]

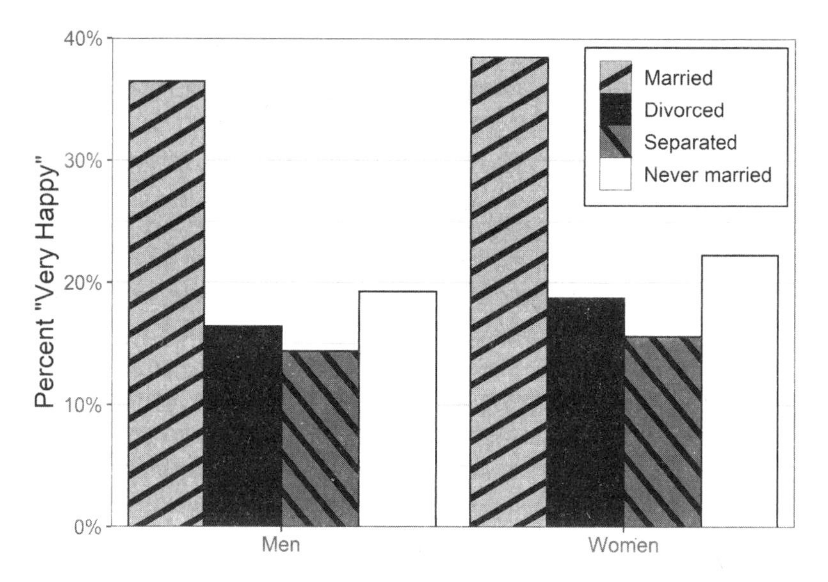

Figure 9.1. Marital status and happiness. Data from 62,226 participants in the General Social Survey, from the National Opinion Research Center, 1972-2021.

It's important to note that, especially in the world of social science, the relationship between marriage and happiness is very strong. Data spanning five decades indicate that among married people, almost 40 percent say they are "very happy," compared to less than 20 percent of those who are single, divorced, or separated (see Figure 9.1). Another analysis that included data from 1972–2008 similarly found that being married was the strongest predictor of happiness, more than work status, age, gender, race, and annual income.[33]

Several studies have also shown that this phenomenon is not limited to Western culture or the United States. When the social scientists Steven Stack and Ross Eshleman studied the correlation between marriage and happiness in seventeen different nations, they found that across all of these cultures and societies, "married persons have a significantly higher level of happiness than persons who are not married."[34]

Evaluating data from over 25,000 people from nineteen different countries, Arne Mastekaasa found that the positive link between being married and happiness was consistent across these different countries and cultures.[35] A more recent analysis of over 36,000 individuals from twenty-seven countries demonstrated the same finding.[36] Overall, the picture is clear across many different cultures and countries: a huge amount of evidence shows that, overall, people are happier when they are married than when not.

But wait a minute. Is it possible that the strong, consistent relationship between marriage and happiness is just an association? In other words, what if married people are happier because they were happier to begin with? Wouldn't it make sense that happier people are more likely to get married in the first place? This is what scientists call a *selection effect*: happy people are *selected* by marriage partners more frequently than those who are not happy.

It turns out that the selection effect may explain some of the benefits of marriage, but it doesn't explain most of them.[37] When sociologists control for measures of health and happiness before and after the initiation of marriage, the data still show that there is something about marriage itself that makes people healthier and happier. Several studies, for instance, have followed people over long periods through the transitions of marriage (and subsequent divorces that sometimes occur). One such study, conducted over a five-year period, found that "young adults who get and stay married . . . have higher levels of well-being than those who remain single."[38] Other research has reached similar conclusions.[39] The scholars Linda Waite and Maggie Gallagher conclude: "The selection of happy and healthy people into marriage cannot explain the big advantage in mental and emotional health husbands and wives enjoy."[40] There just seems to be something about marriage itself, about the commitment between two loving partners, that produces feelings of happiness and well-being.[*41]

It's worth mentioning briefly here that increased happiness is not the only benefit of a stable marriage. Other benefits include better physical health, a lower risk of domestic violence for women when compared to those who are cohabitating, divorced, or separated, a more satisfying sex life, and greater wealth.[42**] As is the case with the marriage-happiness link, a selection effect cannot explain away the association between marriage and these other benefits. In a remarkable way, marriage seems uniquely tailored to human nature to facilitate happiness, health, and well-being.

Certainly, exceptions to this pattern exist. Almost everyone knows someone who has suffered through a terrible marriage. As mentioned above, while there is nothing that will make you happier than a good marriage, nothing will make you more miserable than a bad one. Even

---

* One scholar explains some possible mechanisms whereby this occurs: "Marriage is the greatest source of social support for most people, more than friends or kin, including emotional and material support and companionship. For those who are married, the spouse is involved in and instrumental to a wide range of other satisfactions, including sex and leisure. Being in love is the greatest source of positive emotions. Marriage is good for health partly because it results in better health behavior—married people drink and smoke less, have a better diet, and do what the doctor orders. . . . Marriage is a kind of biological cooperative whose members look after one another and receive mental health benefits, too, as the result of being able to confide in and discuss problems with a sympathetic listener." Other scholars explain: "Having a partner who is committed for better or for worse, in sickness and in health, makes people happier and healthier. The knowledge that someone cares for you and that you have someone who depends on you helps give life meaning and provides a buffer against the inevitable troubles of life."

** Married men earn more than their single counterparts, and, like happiness, the effects cannot be entirely explained by selection factors. For women, the relationship between marriage and money is more complicated, principally due to childbearing.

supposedly happily married couples all have their moments, days, or even seasons of difficulty and disappointment. However, most couples who stay together are happy most of the time. Fortunately, most marriages are relatively good and, overall, you are more likely to be happy when married than when not.

◆

To briefly abridge a huge amount of research, the concepts of happiness, well-being, and satisfaction are complex and multifaceted. They are sometimes defined differently by different experts. There are many factors that influence and affect these emotional states of being, including money, health, and education. But my hope is that, by this point, I have convinced you in some measure that of all these factors, relationships are the most fundamental to human happiness and well-being. So where does the rewarding and satisfaction-inducing characteristic of relationships come from?

On some level, it must come from our evolution. In part, it's most likely an extension of the necessarily deep and loving relationships that human parents have had to develop with their young children to ensure their survival. It's also surely because strong social connections have always served us so well. As noted earlier, our ancestors likely faced a range of unpredictable challenges. Rapidly changing weather, injury, and attacks from wild animals all made life extremely dangerous. Those groups and families who formed strong social connections were more likely to survive. As a consequence, deep relationships became extremely rewarding.[43]

This line of thinking is crucial to what I am trying to argue in this book. Specifically, this is one of the key ideas I hope you'll take away from it: *A great body of evidence demonstrates that our relationships are the most important factor in our happiness and well-being. If evolution is responsible for this, as it seems to be, then the closest of our relationships—our family relationships—should have the most bearing, for better or worse, on our mental health and happiness.*[44] *This is the way we have been psychologically and evolutionarily engineered. This is the way we were created.*

Abundant evidence supports this idea. For instance, in 2019, the Pew Research Foundation conducted a survey about what people find most meaningful and satisfying in life. They asked nearly 19,000 adults from seventeen countries an open-ended question: "What about your life do you currently find meaningful, fulfilling, or satisfying? What keeps you going and why?" After coding and categorizing all the responses, the most common response was (you guessed it): family and children.[45] In a similar Pew survey among Americans, 69 percent of adults mentioned family when describing what provided them with a sense of meaning in life. This was twice as common as the next-most meaningful category: career (mentioned by 34 percent).[46]

For most of us, this is a self-evident truth of the human experience. For me, it's obvious which parts of my day are the most meaningful. It almost always has to do with coming home from work and interacting with my wife and five children. From roughhousing with my young boys to having intelligent conversations with my teenage girls to the rare but precious occasions when I can have a deep conversation with my wife, these moments are full of warmth, love, and meaning.

Happiness and Children

Many so-called experts in mental health have proclaimed that if you want to be happy, it's best to avoid having children. In many studies, childless couples rate slightly higher on measures of subjective well-being compared to couples with children.* How does this make any sense if, as I'm claiming, family is at the root of our happiness?

Well, it turns out that the relationship between having children and happiness is very complicated. Certainly, a lot depends on how we define our terms. In this context, "happiness" might not be the best term to use. It probably makes more sense to describe parenthood as "rewarding" or "meaningful." Certainly, most people do not find having and raising children to be pleasant in the same way that, say, a vacation at the beach might be pleasant. Parenthood is rarely relaxing and is often a high-stress endeavor. "Although parenting has many rewarding moments," the Harvard psychologist Daniel Gilbert has written, "the vast majority of its moments involve dull and selfless service to people who will take decades to become even begrudgingly grateful for what we are doing."[47]

But when researchers ask people what makes their lives rewarding or meaningful, most put children at the top of the list. A Pew Research Foundation study of about 5,000 adults reported that parents were almost twice as likely to say that spending time caring for their children is "very meaningful" compared to spending time at work.[48] Another study found that the more time parents spend caring for their children, the more meaningful they find their lives.[49] Moreover, adults with children generally find life more meaningful than those without children.

Another point worth mentioning is that the supposed link between being a parent and unhappiness is probably due more to cultural and economic factors rather than being an intrinsic and immutable

---

* It's important to point out here that the supposed happiness gap between childless couples and parents is relatively small in magnitude. The difference between childless couples and parents who say they are "very happy" is a few percent. To put this in context, the consistent happiness gap between married and non-married individuals reporting they are "very happy" is approximately 20 percent (see Figure 9.1 above).

characteristic of human nature. Our culture has changed drastically over the last century or two, perhaps in no area more dramatically than that of having and rearing children. Two hundred years ago, parents used to rely on their children to contribute economically to the family fund. Nowadays, most kids are only expected to play with toys, do their homework, and hopefully stay out of trouble.* Instead of contributing to the family fund, children have become consumers of costly goods and services, such as musical instruments (including pricy lessons), expensive video games and electronics, and (most dreadfully) college tuition. Over the last one hundred years or so, the cost of raising children has grown by a huge amount. This cultural shift has unfortunately led too many people to view children as not much more than a huge economic liability.

Some scholars, noting these cultural changes, have concluded that it is the financial stress that comes from parenting that contributes to the slightly lower rates of happiness among parents. The Dartmouth economist David Blanchflower is one of them. When Blanchflower and his colleague Andrew Clark analyzed data representing over a million people, they found that once financial strain was accounted for, parents tended to be happier than their childless peers.[50] "It's not that children make you unhappy," Blanchflower told *The Atlantic* in 2019. "It's the fact that they bring lots of expenses and difficulties. You have to buy the milk and the diapers. And that financial pressure gets muddled up with this."[51]

Another cultural factor that explains the association between being a parent and lower happiness ratings is juggling the responsibilities of

* Don't get me wrong; protecting children from forced labor was a wonderful achievement. But the pendulum may have swung a bit too far to the point that our children don't learn how to work and contribute to the greater good of the family.

family and work. In the United States, despite the fact that we have a much more affluent lifestyle than we did a hundred years ago, adults (especially women but also men) are logging more and more hours working outside the home. In recent decades, many people (especially low wage–earners) have been compelled to spend more of their waking hours working to keep up with the rising cost of living. As a result, any supposed "work-life balance" (such a trendy term in today's corporate world) has become a raging conflict. In countries that have more family-friendly work policies (such as extended and paid maternity/paternity leave), having children can be associated with more happiness, not less.[52] When policy makers, governments, and cultural opinion leaders have been able to help resolve this conflict, people who have children tend to be happier than those who don't.

Another way to think about this issue is to frame it in economic terms. One way to know if someone really enjoys what she bought is to ask whether she would make the same decision and buy it again. If she would not, then she must not have enjoyed it very much. This is buyer's remorse. When parents are asked if they had to do it all over again, would they have children, the overwhelming majority (91 percent) say yes. For parents, buyer's remorse is uncommon. But among childless adults, *non*buyer's remorse is extremely common. When adults over age forty with no children were asked in a survey if they would remain childless if they could do it all over again, more than two-thirds said they would actually choose to have children.[53]

In short, the large body of research examining the relationship between being a parent and happiness/satisfaction is complex. Being a parent is frequently not pleasant, but it is extremely rewarding (at least it has the capacity to be). Most parents reflect that having children requires immense sacrifice, but also provides a profound sense of joy and meaning.[54]

The (Misleading) Quest for Status and Material Possessions

If relationships, especially with our families, are so integral to living a good life, why are we so prone to sabotage them? Why are there so many dysfunctional families?

One of the reasons that we sometimes undermine our relationships comes from the dual potential of human nature. As discussed in prior chapters, we have the tendency to be selfish, aggressive, cruel, and miserly. Such characteristics are clearly damaging to our closest relationships. The pull of selfishness is strong. In so many instances, these instincts put us at odds with those whom we love the most.

Another reason is that Western culture so highly values economic and social status. Many people choose to devote extraordinary amounts of their time to career and work—and far too many others are *forced* to do so in order to make ends meet. It doesn't help that some of us do this because, as a species, we seem to have an insatiable appetite for status and material possessions. As Adam Smith once put it, "The desire of food is limited in every man by the narrow capacity of the human stomach; but the desire of the conveniences and ornaments of building, dress, equipage, and household furniture, seems to have no limit or certain boundary."[55]

If we are not careful, this drive for status and material possessions can hurt our personal and family relationships. While it is true that professional achievement can bring some measure of happiness and well-being, this pales in comparison to the level of satisfaction derived from warm and healthy personal relationships, especially with our family members. "Striving for individual status," the late Harvard psychiatrist Armand Nicholi once wrote, "ultimately proves empty and frustrating if gained at the sacrifice of these relationships."[56]

Our recent cultural revolutions have not helped in our quest for the good life. The 1960s and 1970s ushered in a renewed and singular focus on the individual. Happiness and fulfillment, we were told both by scholars as well as cultural icons, could be found primarily in the expression and satisfaction of personal desires. Family commitments—marriage and children especially—often restricted the fulfillment of these desires. Carl Rogers, one of the most influential psychologists of the late 20th century, declared with stunning confidence: "The only question which matters is, 'Am I living in a way which is deeply satisfying to me, and which truly expresses me?'"[57] Other psychologists and philosophers have been happy to endorse this self-centeredness.[58]

What followed was a rapid decline in our commitment to family and community—and a steady increase in individualism. In the United States, the divorce rate approximately doubled in the latter half of the 20th century. During this same period, the fertility rate dropped by almost 50 percent, a staggering decrease. Indicators of our commitment to the larger community have also demonstrated a decline, with a smaller proportion of people belonging to organizations and attending church, and less of our discretionary income dedicated to philanthropic giving.[59]

And what do we have to show for it as a society? Our income has grown markedly in the last sixty years, and opportunities for education, culture, and leisure are at an all-time high. But have we become any happier as a result? If anything, our happiness has declined. Depression and anxiety have increased markedly. In the last twenty-five years, the United States has seen a surge in suicides. Drug overdose deaths, which some have called slow-motion suicide,[60] are at epidemic levels.

One thing seems certain: the more people seek to achieve personal fulfillment at the expense of relationships, the more social problems they develop and the less happiness they actually realize.[61] This is probably the psychological phenomenon that Jesus was referring to when He said, "Whosoever shall seek . . . his life shall lose it."[62]

To sum up: Our cognitive psychologies were forged during a period in our evolutionary past when resources were scarce. Because of this, our intuitions often lead us to believe that more material possessions will make us happier. We are psychologically wired to believe that if we can just gain a few more material goods, or get that promotion, or win that award we so desperately crave, *then* we will finally be happy. But modern psychological studies have repeatedly made clear that this quest for status and material possession does not lead to enduring happiness or satisfaction. Much of the strife and harm that we inflict on ourselves and those we love comes from this cognitive illusion.

Maximum Freedom Does Not Lead to Maximum Happiness

Another cognitive illusion that often leads us away from achieving enduring happiness is the deceptive but alluring belief that maximizing our freedom will result in maximum personal happiness. This might be referred to as the trap of "keeping your options open." But in the last few decades, modern psychological science has shown that, in some contexts, keeping our options open leads us to be less satisfied, not more.[63]

Consider, as just one example, an influential study of college students and photography conducted in 2002 by Dan Gilbert and Jane Ebert. In this experiment, students were asked to participate in a brief course on photography (this was before the days when everyone walked around with high-quality cameras built into their cell phones). As part of the course, the students ended up with two final, high-quality photographs that were personally meaningful. They were told they could keep one but that the other had to be turned in to the course directors as proof the students had participated in the course. The students could choose which photograph to keep. Half of the students were told that the decision they made was immediately final (the committed group); the other half were told that they could swap photographs several days following their decision (the keep-your-options-open group). All students were later asked to rate how satisfied they were with the photographs they chose. Most people (including those who took part in the study) would probably guess that the students who could keep their options open would be happier with their choice. After all, if they later decided that they didn't like it, they could simply exchange their photo for the other one. But the true outcome was that the students who were committed to their photos were *more* satisfied than the students who kept their options open. Perhaps even more surprising was that even after the chance to swap photos had expired, the students who were committed early to their photos were more satisfied than those who kept their options open.

One of the key conclusions that Gilbert and Ebert made was that *owning* an outcome generally makes someone more satisfied with it. Using an analogy, they explain:

> If every apple contained a fixed amount of sweetness, then decision making would merely be an attempt to estimate that sweetness before deciding which apple to buy, and because such estimates may be mistaken, the opportunity to exchange apples after taking a bite would be an invaluable asset. But the fact is that irrevocably owned apples are sweeter than those that are merely tasted. The ratio of fructose to cellulose is an objective and unchanging property of apples, of course, but the experience of sweetness is a subjective property that increases when *an* apple becomes *my* apple.[64]

In other words, in some contexts, when we commit to a choice, we tend to be more satisfied. Certainly, in many instances, more freedom is better. This is surely the case when we talk about oppressive governments in countries where basic human rights of speech, religion, and assembly are restricted. But in Western democratic societies, the pendulum may have swung too far the other way. At least in some contexts, commitment trumps keeping your options open in terms of leading to happiness and satisfaction.

Now, it may seem like a stretch to say that a controlled experiment involving college students and photography shows how maximum freedom does not equal happiness. But this basic concept seems to be at play in important real-world scenarios. Perhaps one of the most meaningful examples involves the trade-offs between marriage and cohabitation. As the cultural revolution of the 1960s and '70s unfolded, many sociologists were initially surprised to discover that married couples are happier and more satisfied than couples who cohabitate without marrying. This has been found consistently in many studies and across many different cultures.[65] The psychological trap of "keeping your options open," as demonstrated by the photography study, likely explains part of this sociological phenomenon. Whereas marriage is about commitment, cohabitation (in many instances) is clearly about keeping your options open.* If we keep in mind all the psychological

* Commitment really does seem to play a central part in this phenomenon. Cohabitating couples who are engaged to be married when they move in together (a strong sign of commitment) behave more like married couples, whereas cohabitators without an explicit commitment are less happy and satisfied, save for the future less, and have less stable relationships. There are undoubtedly cohabitating couples with stronger commitments than some married couples. But marriage and cohabitation represent different ends of the commitment spectrum, and it appears that an explicit expression of commitment to each other and to the broader community (through marriage) really does provide tangible benefits to the couple.

data that shows how important relationships are to happiness and well-being, the finding that married partners are consistently happier than non-married cohabitating partners shouldn't be surprising. All other things being equal, a committed relationship—wherein partners are married—would lead to a stronger (and happier) relationship than one in which both partners are keeping their options open for other prospects. *New York Times* columnist David Brooks expressed this concept well when he wrote: "People are not better off when they are given maximum personal freedom to do what they want. They're better off when they are enshrouded in commitments that transcend personal choice—commitments to family, God, craft, and country."[66]

How Will We Measure Our Happiness?

I'll mention one final reason that we seem so prone to undermine our relationships (and thus the source of greatest satisfaction): Western culture is obsessed with measuring performance. Francis Galton (an eminent psychologist of the 19th century and cousin of Charles Darwin) once famously said, "Whenever you can, count." Modern society in the Western world seems to have taken his advice to heart. Many people who are successful from a material perspective have adopted this mantra with gusto, and our collective psyche has been shaped to reflect this. Given that we are wired to seek achievement (see Seligman's theory of well-being from above), our jobs and professions are usually the most immediate way to measure whether we are successful. We publish a paper, we send a shipment, we land a contract or a sale. Each of these provides a very tangible and measurable sense of achievement. Each also provides a short-term emotional "high" that is quasi-addicting. Behaviors and outcomes that are easily measured tend to be reinforced.

But the strength of relationships is very hard to measure. And relationships, especially with our children and other family members, often take a long time to mature. Investing time and energy into raising children or deepening our relationship with our spouse often doesn't yield immediate and tangible evidence of accomplishment. As a result, many are drawn to allocate their most precious resources to those activities that produce the most immediate and measurable indications of success. Investing our time and energy into raising children or deepening our relationship with our spouse, the late Harvard professor Clayton Christensen remarked, "often doesn't return clear evidence of success for many years. What this leads us to is over-investing in our careers, and under-investing in our families—starving one of the most important parts of our life of the resources it needs to flourish."[67]

Consider the following thought exercise: Imagine that you find yourself toward the end of the day with an unexpected extra few hours of time. How would you spend it? Would you catch up on a work assignment? Would you read a book to one of your children? Or would you write a love letter to your spouse? If you are like me, when you honestly conduct this exercise, the answers you start to come up with are somewhat disturbing and may not be leading you in the direction of enduring happiness.

◆

For centuries, philosophers and scholars have debated the root of human happiness. But, as we've seen, happiness is a tricky concept. It is often defined differently by different people (and in different studies). As a concept, happiness overlaps in many ways with other emotions, like well-being and satisfaction.

Perhaps the better question is this: How does one live a good life? A good life isn't merely the sum total of highs and lows. It's not the maximization of pleasure and the total avoidance of difficult or painful experiences. It involves living in a way that is rewarding, meaningful, and full of purpose; this often requires giving up something good now for something better in the future.

When we ask ourselves what makes for a good life, most of us answer by pointing to our families. This seems to be a consequence of the way we evolved. It is how we were created.

Chapter Ten

The Good Society

In the late 1840s, John Humphrey Noyes established the Oneida Community, one of the longest-lasting attempts at communal living in the history of the United States. Noyes was a preacher who believed that communal perfection could be attained in this life. Born in Vermont in 1811, Noyes studied theology at Yale after a series of personal religious experiences. In some ways, his ideas were remarkably progressive; he advocated for a more egalitarian approach to a woman's role in society, and he helped organize one of the first antislavery societies in the United States. But in other ways, his approach to life during his time at Yale became increasingly radical, bordering on anarchy and insurrectionism. ("I have renounced active cooperation with [the U.S. Government] the oppressor on whose territories I now live"; "My hope of the millennium begins . . . at the overthrow of this nation."[1]) As a result of his ever more bizarre and radical ideas (including the belief that he did not sin[2]), his professors branded him a heretic and he was expelled from the Yale Divinity School.[3]

Undaunted (perhaps even emboldened[4]), Noyes returned to Vermont and set out to start a new social order. In so doing, he sought to undermine marital and family relationships such that "loyalty was raised to the level of the entire community."[5] Judging that his "new relationship to God canceled out his obligation to obey traditional moral standards,"[6] Noyes introduced his followers to what he called "complex marriage." Through this arrangement, every male in the community was considered linked in divinely sanctioned matrimony to every female. Individuals were thus encouraged to exchange sexual partners frequently, so long as both adults consented. Birth control was also readily encouraged and enforced culturally through the self-discipline of males. This controversial approach to family life soon created tensions with the community's Vermont neighbors, so Noyes and his followers relocated to Oneida, about ten miles east of Syracuse in upstate New York.

In New York, Noyes worked hard to win converts. Eventually, the Oneida Community swelled to about 300 individuals, with satellite communities established in Wallingford, Connecticut; Newark, New Jersey; and Brooklyn, New York.

As the community grew, Noyes refined his system of complex marriage and developed a mechanism for introducing new initiates into the social order. As teenagers neared maturity, initial partners were chosen for them who were often much older and more experienced. Those who were less committed to the religious order were often paired up with more devout individuals, in hopes that their commitment would deepen. Sexual exchanges were tightly regulated and even recorded; each required preapproval by Noyes himself or another of the leading elders in the community.

Individuals could request to have children, but the community leaders required that children live in separate quarters from their parents and be reared communally. Decisions regarding procreation and child-rearing were made by formal committees. All of these practices derived from Noyes's belief that man's natural commitment to his family was antithetical toward the building of a strong community. By weakening the natural bond between adult and offspring, Noyes sought to strengthen everybody's commitment to the greater social order.

The upending of the natural human tendencies embedded within the nuclear family proved difficult to enforce culturally. Jealousies invariably arose, and even Noyes himself was not immune from this. Young people, in particular, "often found the thought of sharing their sweethearts unbearable."[7] One sixteen-year-old boy threatened suicide at the thought of his love being with someone else.[8] There was a need for constant preaching against the "evils" of exclusivity, or what was termed "idolatrous love." Noyes developed a powerful culture to keep behavior in line with his vision for communal living. Couples who drifted toward monogamy were disciplined publicly. Invariably, some young people escaped from the community, driven in large part by the desire to have a monogamous sexual relationship and not to be linked with partners who were often a generation older than them.

The arrangement of communal child-rearing and the practice that children lived apart from their biological parents were also a source of tension and dissatisfaction, especially for women. (Noyes considered the possessiveness of mothers toward their own children an impediment to communal happiness.) One journal tells of a mother who, taking advantage of a rare private moment with her son, broke down and begged to know, "Darling, do you love me?" The son recalled later of this moment, "I always melted. My marbles and blocks were forgotten. I would reach up and put my arms about her neck. I remember how tightly she held me . . . as though she would never let me go."[9]

Another mother wrote of how again and again, she gave in to the "mother spirit" that "begets selfishness and idolatry"[10] by visiting her young daughter more frequently than she was allowed. As a punishment, she was forbidden from seeing her daughter for days or even weeks at a time. Her daughter later wrote how, upon catching a glimpse of her mother after a two-week period of separation, she "rushed after her screaming." Her mother, recognizing that a breach in communal rules would result in a prolonging of the sentence of separation between the two, quickly tried to escape from her young daughter. But the daughter "rushed after her, flung myself upon her, clutching her around the knees, crying and begging her not to leave me, until some [community leader], hearing the commotion, came and carried me away."[11] The records suggest that incidents such as this were not uncommon.

Despite this strained form of family life, the Oneida communalists became extremely industrious. Like many Utopian experiments of the time period, the community abolished the notion of individual property rights in favor of communal property ownership. They developed a culture that prized work as a pleasure and honorable duty. As a result, they successfully established a variety of business endeavors, one of which survives today and is among the world's largest manufacturers of stainless-steel silverware.

The experiment came to an abrupt end in 1879, some thirty years after its initiation in rural Vermont. In the middle of a summer night, Noyes fled the community for Canada; he had failed to transition the leadership of the community to his son and was alarmed at the growing threat of legal prosecution at the hands of neighbors who were outraged by his unconventional teachings. His escape was so secretive that it was unknown to virtually all but a few of his most trusted followers. Without the charismatic and strong-willed leadership of Noyes, the culture that he had established evaporated almost overnight. Within two months of Noyes's furtive departure to Canada, the practices of complex marriage and communal child-rearing abruptly dissolved. While some of the business endeavors of these communalists continued for some time following Noyes's escape, the unconventional social order that made Oneida a unique communitarian experiment was gone.

The previous chapter focused on what makes for a *good life*. This is a complex question. Wealth, health, education, and many other factors all play into this. But the most important factor is relationships. As I have argued, because evolution has shaped our natures, those people with whom we share close family ties are the most relevant to biology and evolution. We could therefore expect these familial relationships to be eminently important to achieving the good life.

Now let's pivot and focus on how this is related to what makes for a *good society*. Sigmund Freud argued that the good life and the good society were incompatible.*[12] Fortunately, he was wrong. Just as strong family relationships are the most important factor to a good life, a wealth of sociological data demonstrate that relationships are also vital for a good society. Helping people forge stronger relationships, especially with their families, will not only help them be happier but will also help them be better—that is, help them choose altruism over selfishness, cooperation over aggression, generosity over greed, honesty over dishonesty, and so forth. As I'll explain, this is most apparent in helping men strengthen their relationships with their children.

But first, what do I mean by a "good society"? What does this entail?

This is a difficult question to answer. Most of us have at least a vague sense of what a good society would look like. It would provide

* Freud concluded that humans, with what he deemed were insatiable appetites and passions, would ever be an enemy to organized society: "Much of the blame for our misery lies with what we call our civilization, and we should be far happier if we were to abandon it and revert to primitive conditions. . . . If civilization imposes such great sacrifices not only on man's sexuality, but also on his aggressivity, we . . . understand why it is so hard for him to feel happy in it. Primitive man was actually better off, because his drives were not restricted. . . . Civilized man has traded in a portion of his chances of happiness for a certain measure of security."

protection to the vulnerable. It would value principles of truth and justice. It would afford maximum opportunities for individuals to obtain happiness. A good society would uphold people's ability to choose, so long as their choices did not cause harm to or interfere with the rights of others. Other important principles would include security, opportunity for advancement, privacy, safety, social inclusion, and so forth.

The problem is that if we listed ten aspects of a good society and asked thirty people to list them in order of importance, we would likely get thirty different lists. Much of the polarization of American politics is due to the debate over the ordering of these principles. But in my opinion, much of the political debate overlooks another, more fundamental problem that prevents us from achieving a good society. This problem is that the darker propensities of human nature—selfishness, greed, dishonesty, aggression, and so forth—are at odds with communal well-being. Many respectable and intelligent people disagree on the order of importance of principles of the good society. However, virtually everyone would agree that if there were a way to make people more virtuous, then achieving the good society (even though we may not agree on the precise definitions of this) would be much easier.[*13]

Much has been written about achieving a good society. In essence, this has been one of the principal goals of political theory since its origins. Like the Oneida Community, there have been many groups that

* One political commentator, noting this principle, remarked: "The chief illusion of modern political activity is the belief that you can build a system so perfect that the people in it do not have to be good. . . . A society is a system of relationships. If there is no trust as the foundations of society, if there is no goodness, care, or faithfulness, relationships crumble, and the market and the state crash to pieces. If there are no shared norms of right and wrong, no sense of common attachments, then the people in the market and the state will rip one another to shreds as they vie for power and money. Society and culture are . . . more important than politics or the market. The health of society depends on voluntary unselfishness."

have deliberately abandoned their surrounding societal structure and drastically remade social rules in an attempt to create a utopia. Some of these communities were united by deeply held religious beliefs; others were entirely secular. Many such attempts were well intentioned, with their founders recognizing great social injustices that existed in the status quo arrangement. Others were led by colorful and idiosyncratic zealots whose growing thirst for power corrupted their motives. In the early 19th century, with the American spirit of independence fresh in the public mind, several dozen Utopian experiments attempted to radically refashion social arrangements and mores. In addition to the Oneida Community, these efforts included Brook Farm, the Shaker community, the Raritan Bay Colony, the New Philadelphia Colony, the Harmony Society, and the Free Lovers at Davis House. Many of these communal experiments began with praiseworthy aims. Many were progressive in their treatment of women. And many of them abandoned the notion of private property in favor of communal ownership. But all of them utterly failed, and most within a year or so. (The Oneida group was one of the rare exceptions that lasted more than a few years.)[14]

It's unlikely that there is only one reason for the failure of all of them. However, a common theme of many of these communal experiments was that adherents were required to abandon their close family relationships. Related to this, many groups drastically altered the approach to sex. On the one hand, some groups (like the Oneida) encouraged liberal sexual interaction among its members. These communities tried to suppress the natural desire for exclusivity between partners. On the other hand, groups like the New Philadelphia Colony, the Harmony Society, and the Shakers advocated for complete abstinence. The thinking of these community leaders was that by weakening the bonds between spouses, people would naturally develop stronger bonds to the community. This is also why some communities tried to weaken the bonds between parents and children by forcing them to live separately from each other.[15] But, as demonstrated by the Oneida Community, this approach was misguided and backfired.*

As we have touched upon already, kin selection is the root of our deepest and most biologically driven forms of love, trust, loyalty, kindness, and many of our other positive attributes. We are biologically primed to love and care for our families, more than any other relationships. This is a consequence of our evolution. Societies that try to subvert these deepest forms of love are not sustainable. By extension of this logic, societies that nurture and support healthy family relationships will reap great communal benefits. This occurs in at least two critical ways: 1) by helping men redirect aggressive tendencies toward prosocial ends and 2) by improving outcomes for children.

Fatherhood: Helping Men Become Better

Given that so many of the deepest capacities of our better natures are the result of how evolution shaped our family relationships (i.e., kin selection), it would logically follow that when adults are actively engaged in rearing their children—and when these family relationships flourish—their better natures are more likely to predominate. As it turns out, a vast amount of sociological data supports this theory. Family life, in essence, is nature's most powerful way of helping us choose our better natures. This is especially relevant for men, who are responsible for the majority of social problems in most cultures. Compared to women, men are almost five times more likely to commit

* The authors of communism also advocated for an abandonment of the intact biological family structure, which they saw as contributing to oppression and inequality.

arson, seven times more likely to steal, ten times more likely to murder, and over eighty times more likely to commit rape.[16] More than 90 percent of all prison inmates are male.[17]

But when men "settle down" and become fathers, they often experience a subtle but significant change in nature. Fatherhood seems to channel male energy and aggression toward constructive and prosocial ends. Indeed, across cultures, becoming a father has been observed to lead men to become less selfish and more socially responsible.[18]

Of course, it's more than just having children that leads to increases in prosocial behavior. Men also need to be engaged in rearing their children for these prosocial behaviors to be realized. But unfortunately, compared to women, men have less of a biological tie to their children. This is because men contribute comparatively little energy and effort to the physical conception of a new human being. From a purely biological standpoint, a man can contribute his part to the conception of a child in a matter of minutes. There are even instances where a man conceives a child without later knowing it; for a woman, this scenario is impossible. Women have an enormous biological investment in their offspring. A woman carries an unborn baby for the better part of a year; following birth, a new mother further strengthens her biological bond with the baby through nursing. The way the human race has evolved has caused an imbalance in the strength of biological ties between men and women and their children. Due to their substantial biological investment to child-rearing, women naturally have a strong emotional tie to their children. In contrast, men have a fairly fragile link. In almost all cultures, men are more likely than women to be uncommitted and detached parents. A strong social and cultural commitment is needed to link men to their children and enhance their role as fathers. For most of history, the institution of marriage has served this express purpose.

Only since the drastic rise in divorce that occurred in the latter part of the 20th century have scholars fully realized how important a role marriage plays in linking men to their children. The most obvious ways that marriage links a man to his children is by having him live with their mother. But when parents divorce, they live separately. Since the beginning of the 20th century, children of divorced parents have mostly lived with their mothers, even in cases of so-called joint custody.[19] At this point, the relationship with dad is often reduced to phone calls, weekend visits, and child support payments.[20] As the journalist Solomon Jones writes from personal experience, divorced fatherhood is a "disjointed tapestry of love and distance, longing and hurt." Jones goes on to explain how living separately from his child "was painful because a father's love is so often expressed through providing and protecting. And it's difficult to provide and protect without presence." "Fatherhood," he concludes, "works best when it is paired with motherhood and sealed by marriage."[21] The data shows that Jones's experience is typical of divorced fathers, who spend much less time with their children than married fathers.*[22] In general, if a man's marriage declines, the quality of his fathering also declines.[23] In fact, one sociologist commented that trying to gather fertility data from males is "notoriously unreliable because many men simply 'forget' children" who are not living with them.[24]

A wealth of sociological data shows that, especially for men, marriage and parenting are closely connected.[25] Some researchers even refer to them as a "package deal."[26] Indeed, it has long been observed that a strong marriage is a very strong predictor of how involved a man is in the lives of his children.[27] This package deal of marriage and fatherhood

* About 65 percent of children from divorced parents report a poor relationship with their father, compared to only 29 percent of children from families where mom and dad remain married.

seems to civilize and direct the male nature toward prosocial ends. As one sociologist notes:

> Family life—marriage and child-rearing—is an extremely important civilizing force for men. Control of the more unseemly male passions is a requisite for successful family living. Family men must develop those habits of character, including prudence, cooperativeness, honesty, trust, and self-sacrifice, that can lead to achievement as an economic provider. In addition, marriage notably focuses male sexual energy, and having children typically impresses on men the importance of setting a good example.[28]

Interestingly, even without children in the picture, marriage has strong prosocial effects on men. However, marriage and fatherhood seem to be inextricably linked. For instance, many men readily admit that they will give up their bad habits when they get married. They presume that with marriage will come the commitments of fatherhood and the duty to be a good example to any children born within the union of that marriage. The sociological data suggest that, by most measures, men generally keep this promise of taming their behavior when they marry.[29]

For example, in virtually all measurable ways, married men behave better than men who are unattached and devoid of family commitments.* Certainly, there are many men who remain unmarried and also make outstanding contributions to society. But on average, married men are less likely to behave in criminal and violent ways than men who are not tied to family obligations.[30] They are less likely to suffer

---

* They are also better off in terms of health, financial status, and happiness compared to unmarried men.

from disease and disability (because they drink and smoke less) and less likely to end up incarcerated.[31]

Unfortunately, the retreat from marriage and family life that occurred in the latter half of the 20th century was accompanied by a marked increase in crime, incarceration, depression, and other forms of mental illness.* As one scholar notes: "Although the rise of contemporary fatherlessness is associated with the revolution of self-fulfillment and its promise of greater adult happiness, no one, least of all men, seems to be any happier as a result."[32]

Cultural factors were not the only reasons for the retreat from marriage and family life in the latter half of the 20th century. Economic factors likely also played an important role. As many blue-collar jobs disappeared, it became harder for men to support their families. Research has shown that improved labor markets increase the probability of marriage for men.[33] But marriage also plays a critical role in linking men to their children and helping them mature.

What is so special about marriage and fatherhood? How do these institutions help men become better? There are several likely ways this happens. For one, marriage and family life can provide powerful motivation for men to be actively engaged in prosocial endeavors, including gainful employment. Doug Taulbee exemplifies this. At the age of eighteen, Doug was working in a factory in Indiana, living in his parents' basement and earning little more than minimum wage. "I really didn't have a care in the world," Doug recalled. "I didn't really have any bills." He was content in his life and didn't have any reason to

* It's true that violent crime dropped dramatically in the 1990s, without a strong return to the norm of intact families. But this came at great cost. Incarceration rates increased about 500 percent from 1970 to 2000, and they haven't really changed dramatically since. In addition to this drastic increase in incarceration rates, many scholars think a combination of increased police force, more innovative policing strategies, tougher gun control laws, and other factors were related to the decrease in crime rates observed in the 1990s.

aspire to anything more. That is until he got married. Upon marrying at the age of nineteen, he developed powerful motivation to improve his economic outlook. Having children only enhanced this drive. "I had to step up and think about something besides myself and start taking care of them." Doug joined the army, quit his factory job, and his income increased by 50 percent. He was invested in seeking better employment opportunities to provide for his family. Following his stint in the army, he became a finance manager at a car dealership, where his income quadrupled compared to his army job.[34]

The story of Doug Taulbee is typical. It is representative of what many sociologists have recognized for decades: marriage has the potential to transform adult behavior. For men, more than any other life event, marriage provides an intrinsic motivation to provide and protect. (This motivation is often magnified once children are born into the marriage.) Men who are married work harder and more strategically than their single peers. As a result, in the United States, they earn more than their single counterparts.[35] Married men have better health and engage in less risky behaviors, like drug use or heavy drinking.[36] They are less likely to commit crimes of any type than single men.[37]

A Note About Selection Effects

Of course, the skeptic may protest that the supposed advantages of marriage are merely explained because the good men are the marrying type. After all, why would any woman want to marry someone with violent or criminal tendencies? Recall from Chapter 9 that this is what social scientists refer to as a "selection effect." The men who are more motivated and prosocial to begin with are more likely to be "selected" into marriage. Hence, any association between marriage and positive outcomes (such as improved health, less risky behavior, higher salary, more prosocial behavior) is not a result of marriage, per se, but only a reflection of the fact that these are the types of men that women would actually want to marry in the first place. While there may be some truth to this, over time, the vast majority of research suggests that only a portion of the marriage premium (less than half by some estimates)[38] is due to this "selection effect." In other words, there seems to be something about marriage itself that transforms men and helps them mature. When controlling for differences in education, ethnicity, regional economic factors, and intelligence, married men still outperform their single peers in a host of social indices.[39] Also, studies that follow the same men before and after marriage still show that marriage provides a strong intrinsic motivation for men to mature, live healthier, and work harder and more strategically.[40] Additional research has found that married men earn 26 percent more than their identical twins who are single.[41] When married men divorce, many of the benefits of marriage also erode.[42]

Like marriage, becoming a father also generates powerful motivation for men to provide for their families. But as I mentioned above, simply conceiving a child doesn't cut it. A man must be active in his social role as a father to realize this motivation. The social worker Charles Ballard recognized this as he helped unwed, absentee fathers reengage with their children in the tough inner-city neighborhoods of Cleveland. Traditional thinking in sociology dictates that the way to help such men is to provide economic security for them: these men must first have jobs before they are able to meaningfully reengage with their children.[43] But Ballard and his colleagues flipped that logic on its head; they found that when they focused on reuniting these men with their children, the men were able to find the intrinsic motivation to improve their own situations. From 1982 to 1995, through a combination of home visits, therapy sessions, parenting classes, and other approaches, Ballard helped reunite over 2,000 absentee fathers with their children. When they entered the program, only 12 percent of the men had full-time employment. But after reengaging with their children, 62 percent had found full-time work, with an additional 12 percent who had found part-time work. Over 95 percent of these men started contributing financially to the support of their children.[44]

This finding would be in line with what social psychologists have long known: intrinsic motivation is much more powerful than extrinsic motivation. From an evolutionary perspective, there is perhaps no more powerful intrinsic motivating force than the duty to provide for your own biological offspring. Society's message to these men has generally been that they should get jobs and be contributing members of their community. But from the perspective of the men themselves, this was comparatively weak motivation and few of them internalized this message. However, once these men were reminded on a regular basis of their paternal (and evolutionary) role, they developed the intrinsic motivation to improve their situation.

The fatherhood initiative led by Ballard is a wonderful approach that yielded strong results. More social initiatives should take note of the powerful intrinsic motivation that comes from helping men fully engage with their children. Yet social programs such as these are extremely expensive. Our enthusiasm for such programs is further tempered when we recognize that this initiative was essentially trying to serve a purpose that marriage has naturally served for centuries in almost all cultures: tying men to their social roles as fathers of their biological children. It would likely have been better for all parties if these men had simply married the mothers of their children to begin with.[45]

There is another way in which marriage and fatherhood help men become more prosocial. When men are engaged in their role as fathers, they connect not only to their children but also to the broader community. Engaged fathers volunteer to help coach their children's softball and soccer teams. In so doing, they get to know other parents and strengthen the social fabric of their neighborhood. They have a vested interest in making the communities in which they rear their children as safe as possible. As one scholar concluded: "Fathering functions to bond an adult and child, but it also functions to bond a father to the larger community. . . . When men enter the child-rearing years, they also become more involved in neighborhood activities and community organizations through their children."[46]

Our understanding of the importance of marriage and fatherhood in helping men to behave in prosocial ways provides a strong impetus to use the levers of government and other measures to encourage men to become responsible husbands and fathers. At the same time, this does not require us to assume a posture of judgment toward men who either fail to succeed on this path or choose another. There are many men who will remain single or childless and nonetheless provide significant contributions to society. Recognizing that our world as a whole needs more married men and fathers does not mean that every man will—or should—marry and become a father.

◆

Becoming a parent is a transformative experience that compels adults to adopt new and more altruistic attitudes. Given their strong biological ties to the infant, women generally undergo this change more naturally. Men need strong cultural sanctions to ensure that this transformation occurs.*[47] As anyone who has done it can affirm, being

* The anthropologist Margaret Mead once noted that there is no society on earth where men will stay married long unless culturally required to do so.

a parent is extremely challenging. It compels us to overcome (or at least frequently put aside) any selfish interests to attend to the material and emotional needs of our young children. In essence, becoming a parent is a crash course in how not to be self-centered. Engaged fatherhood accelerates a man's maturity and enhances his ability to care for and think about people other than himself. In a sense, fatherhood is nature's most powerful way to nudge men to choose their better qualities: altruism, patience, kindness, and concern for others. As the author David Blankenhorn describes it, "Fatherhood, more than any other male activity, helps men to become good men: more likely to obey the law, to be good citizens, and to think about the needs of others. Put more abstractly, fatherhood bends maleness—in particular, male aggression—toward prosocial purposes."[48]

Engaged Fathers Significantly Improve Outcomes for Children

In addition to helping men shape their nature for the better, supporting healthy family relationships will also yield great communal benefits by improving outcomes for the next generation. As scholars have learned during the decades since the sexual revolution, when men are engaged fathers, their children do much better.

One of the pioneering scholars in this area was the Rutgers sociology professor David Popenoe. For several decades, it was rare to read a high-profile news media article about the importance of fatherhood without some reference to Popenoe.[49] Starting his career in urban planning, he was disappointed to learn how much politicking was involved in the execution of urban development. He also came to realize that the behavior of the people inside the buildings had far greater impact on society than the buildings themselves. So he shifted his area of inquiry to the study of marriage and fatherhood and became one of the field's foremost authorities, publishing half a dozen books and over a hundred peer-reviewed articles about these topics.

Popenoe's principal message for decades was that in Western culture, the decline of marriage and the subsequent separation of men from their social role as fathers was a root cause of our most serious social problems.*[50] Popenoe found himself attempting to play the role of objective academic in the midst of a fierce culture war featuring changing attitudes about the role of women, sex, and marriage in Western culture. But his message was unmistakable. Despite the difficulties of sociological research, there is perhaps no other topic in which the data is so convincingly clear. In terms of child outcomes, the best arrangement for child-rearing is that of married biological parents. As Popenoe stated in 2005:

> In my many years as a sociologist, I have found few other bodies of evidence that lean so much in one direction as this one: on the whole, two parents—a father and a mother—are better for a child than one parent. There are, to be sure, many factors that complicate this simple proposition. We all know of a two-parent family that is truly dysfunctional. . . . A child can certainly be raised to a fulfilling adulthood by one loving parent who is wholly

* Popenoe was not alone in this regard. A bipartisan national commission in the 1990s concluded: "Many of the most pressing problems of our cities, and of our nation, are substantially attributable to the dissolution of the family. The overwhelming weight of the evidence compels [us] to conclude that the stable, loving, two-parent home is the ideal environment for children and the strongest possible foundation for long-term societal success."

> devoted to the child's well-being. But such exceptions do not invalidate the rule any more than the fact that some three-pack-a-day smokers live to a ripe old age casts doubt on the dangers of cigarettes.[51]

To help bring attention to this issue, he founded the National Marriage Project in the late 1990s, an academic endeavor to study how marriage relationships form and are maintained.[52] One of the key problems that Popenoe highlighted in Western culture was the increasing prevalence of what could be called volitional fatherlessness, which occurs when men willfully leave their families (as opposed to unwillfully leaving their families through death). But volitional fatherlessness was not always considered a problem among scholars. In the early wake of the sexual revolution, some considered that fathers were expendable. For most of human history, a substantial portion of children grew up without fathers. For example, in the 1870s, about 15 percent of all children in the United States lost their fathers by the age of fifteen.[53]

Prior to the 1960s, the principal cause of fatherlessness was death by war or disease. When divorce and childbirth outside of marriage started to grow in prevalence, many argued that the outcomes for children would be similar to losing a father by death. After all, they reasoned, kids are resilient. Many children had successfully overcome adversity in past generations and thrived; surely, they could do it in the face of divorce or separation. But it turns out that this was not the case. Losing a father to death is a different adversity than volitional fatherlessness.

When children lose a father to death, their mother might tell her children uplifting stories about him, encourage the children to honor his memory and to live up to the ideals he embodied in life. In the child's mind, the father remains revered. But divorce is usually different. The mother often holds resentment toward the father for whatever problems led to the divorce. When she talks about the children's father in front of them, it is often in critical tones. In cases where the woman leaves the man, children can be left to wonder what they did to drive their parents apart. Maybe if they were better children, their parents wouldn't have divorced.[54] These and other similar narratives likely lead to the negative outcomes that are sometimes experienced by children without engaged fathers.

And these differences are fairly stark.[55] Compared to children from divorced or single-parent families, children from intact families with married parents are less likely to experience depression, anxiety, attention-deficit disorder, or have other behavioral problems.[56] Girls from intact families are less likely to attempt suicide.[57] In terms of academic performance, children from intact families are also at a marked advantage. They are less likely to drop out of high school,[58] less likely to be expelled or have to repeat a grade,[59] and less likely to be without a job or school in early adulthood.[60]

It should not be surprising to learn that children without a stable, two-parent family are more likely to experience poverty. In marriage, there is an economy of scale as parents often divide aspects of family labor and specialize. In contrast, when parents divorce (or when they never marry in the first place and end up living separately), it becomes extremely costly to maintain two households. In nearly every respect, single-parent homes are worse off economically than the intact two-parent family.[61] In fact, one sociologist noted that "the vast majority of children who are raised entirely in a two-parent home will never be poor during childhood. By contrast, the vast majority of children who spend time in a single-parent home will experience poverty."[62]

Boys who grow up in households where the biological father is absent are also more likely to become engaged in criminal activity. Also, this seems to be true of criminal activity of a violent nature.[63]

This is not merely a phenomenon of Western culture. One sociologist, citing cross-cultural evidence, concludes that societies in which raising children is considered the exclusive role of women have higher rates of crimes against people and property.[64] When the biological father is not involved in his sons' lives, these boys are much more likely to become involved with illicit drugs and gangs, to commit violent crimes (including those against women), to develop alcohol problems, and to drop out of school.

For females, one of the significant downsides of fatherlessness is an increased risk of teenage pregnancy. Girls who grow up with a disjointed (or no) relationship with their fathers are more likely to become pregnant as teenagers.[65]

The benefits of an intact family can also be experienced by the local community. In other words, family structure affects not only the children in the family, but also the children in the family's neighborhood. The Harvard economist Raj Chetty has conducted analyses that conclude that on the community level, the best predictor of upward mobility—which he defines as the ability to climb the ladder from poverty to wealth in one generation—is family structure.[66] In other words, those children who live in communities with lots of intact families are more likely to climb out of poverty compared to communities with a smaller proportion of intact families. In fact, family structure is an even better predictor of upward mobility than the proportion of adults in a community who are college educated.[67] This finding holds even when accounting for a child's individual family situation. In other words, a child of a single parent who grows up in a community with mostly two-parent families does better than a child of a single parent in a community with fewer two-parent families.[68]

It's important to note here that there are millions of children from single-parent homes who have good outcomes. In fact, by many estimates most children from single-parent homes reach adulthood without significant issues. But on average, children with parents who are married have a significantly higher chance of growing up to be healthy and happy than those whose parents divorce, separate, or never marry in the first place.

So why is it that the intact biological family structure* serves children so much better than other forms? Much of it likely has to do with our evolution. All else being equal, no one is more invested in the success and future development of children than the biological parents.[69] Parents see in their children an extension of themselves and, whether consciously or subconsciously, a propagation of their genetic material. Hence, the strongest and most natural forms of love, altruism, and benevolence are experienced by parents toward their children. This is a consequence of how we evolved (an important aspect of kin selection).

Variations on this evolutionary phenomenon can provide explanations for many of the advantages experienced by children from homes with two biological parents. Consider child abuse. As divorce grew in prevalence, some scholars thought that abuse might be one aspect where not having a father around might actually be beneficial.[70] Given that men are much more violent than women, one might be tempted to think that children in families headed by a single woman would be less prone to abuse. However, it turns out that the opposite is true. One of the greatest risks for child abuse is family disruption, most frequently in the form of divorce and subsequent limited involvement

---

* In this chapter, I don't make any distinction as to how involved extended family is when I discuss the effect that intact families have on children. Some have used the term "modern nuclear family" to denote an arrangement in which parents and their children do not receive significant support from extended family. This is a relatively recent phenomenon in our history that has emerged as geographic mobility has increased.

from the biological father.[71] Many studies have shown that stepfathers are much more likely to abuse children than biological ones.[72] Yet perhaps an even greater danger to children than stepfathers comes from the mother's boyfriend,[73] who has no biological stake in the mother's children and has not made an explicit commitment to her (or the community) in the form of marriage. Biological fathers, it seems, have a strong evolutionary motivation to seek the best interest of their children; this is manifest by the lower rate of child abuse by biological fathers as well as the fact that fathers can serve to protect their children from abuse at the hands of other adults. This is consistent with another finding from research that the less confidence a man has that children are his biological offspring, the more likely the chance that he will abuse them.[74] Overall, children who have a biological father engaged in their lives are less likely to experience abuse than children with a less-involved biological father.[75]

Let's be clear. These issues are not easy to talk about. All of us have friends or loved ones who are divorced or—through some other set of circumstances—find themselves raising children outside of marriage. In many cases, people find themselves in these situations through no fault of their own. The point here is not to send divorced parents on an all-expense-paid Freudian guilt trip. But neither can we simply ignore the data and pretend that family structure doesn't matter when it so clearly does.* Society ought to strike a delicate balance by crafting policies and attitudes that help the less fortunate while also encouraging adults to get and stay married. The research is quite clear: children—and hence the society of tomorrow—are better off when they come from intact families.

* We should also be aware that unstable family structures tend to be most common for children who already face other forms of disadvantage.

◆

This chapter explores one aspect of what makes a good society. This is an extremely complex question, and many factors deserve attention. But if there were a way to make people better, to make them less selfish, more honest, more concerned for others, then achieving the good society would be a more attainable goal. One of the most important factors in this area is marriage and family life. When men are actively engaged in family life, linked by marriage to their role as fathers, they behave better, are more unselfish, less violent, and more prosocial in a host of ways. The outcomes for women and children are also generally much better than in alternative living arrangements.[76] Certainly, there are important exceptions and nuances to this sociological phenomenon. Marriage and family life are not the panacea to all social ills. Clearly, circumstances exist where divorce is justified. But in many important respects, marriage and family life are extremely important goals in achieving both the good life and the good society. This seems to be the way that evolution has shaped our natures.

Chapter Eleven

Final Thoughts

In the opening chapter of his Pulitzer Prize–winning book, *On Human Nature*, Edward O. Wilson emphatically declares:

> If humankind evolved by Darwinian natural selection, genetic chance and environmental necessity, not God, made the species. Deity can still be sought in the origin of the ultimate units of matter, in quarks and electron shells . . . but not in the origin of species. However much we embellish that stark conclusion with metaphor and imagery, it remains the philosophical legacy of the last century of scientific research.[1]

Such has been the thinking of many scientists. Evolution, they would argue, explains away any evidence for God. Yet I come to a very different conclusion. As reviewed in Chapter 2, the emerging picture in biology suggests that evolution was not a random process. Instead, the evolution of life was constrained and guided by higher-order principles, leading many different lineages to develop the same structures and functions. The emergence of life was not an accident, but was seemingly inevitable given the inherent laws of nature. This view of life allows for the possibility that our existence has an overarching purpose.

As I explained in Chapters 3 through 6, the mechanisms of our evolution have left us with competing dispositions, or the *dual potential of human nature*. We are pulled in different directions: selfishness and altruism, aggression and cooperation, lust and love. When we couple this with the finding that we possess a measure of free will (Chapter 7), all this strongly implies that there is a purpose to our existence. This purpose, at least one of them, is to choose between the good and evil impulses that nature has created within us. This life is a test. This is a truth, as old as history it seems, that has been espoused by so many of the world's religions. In my view, these aspects of human nature, including how evolution shaped us, are strong evidence *for* the existence of a God, not against it.

Another purpose of life is, where possible, to form families. Marriage specifically, and family relationships more generally, can be the source of some of the richest feelings of satisfaction and joy. Marriage and family are critical not only to the healthy development of future generations, but also to the continuing maturation of adults.

Rearing children is incredibly hard work. It requires an immense amount of sacrifice and dedication. But parenthood helps adults, and men especially, to become less selfish, to avoid crime and drug use, to become more connected to the community, and it provides them with a sense of meaning that they did not previously fathom was possible. As the family scholars Brad Wilcox and Kathleen Kline state, "Despite the challenges, parenthood remains one of the most transformative and meaningful events in our lives. Our children ground us and enliven us. They give us joy and satisfaction that we cannot imagine having lived without."[2]

This unique balance between the sacrifice and joy inherent in parenthood is the result of our evolution. If God really intends for us to multiply, to replenish the earth, and to have joy in our posterity, then it's no wonder that a process such as evolution was used in our creation.

Closely related to this is *meaning*. What is the meaning of life? Based on this view, it would seem that one such meaning is to develop deep and abiding relationships. At least that's what most people report are the most meaningful aspects of their lives. And our relationships that are most meaningful are those with our families, those with whom we share our genes. This is a function of our evolution. It is how we were created.

To me, this signifies a grander plan for our existence, only a portion of which is visible in our current state. If life is a test, then it wouldn't make sense if there were nothing that followed. It's difficult to believe that our most cherished relationships will cease to exist upon death. Most people intuitively seem to believe this. As of 2021, about two-thirds of American adults reported believing in an afterlife where they would be reunited with loved ones.[3] The strong attachments that derive from our family relationships represent the deepest forms of love, tenderness, and friendship that God has created in the flesh. Surely God has provided some way for these greatest forms of affection to endure. This is one of our most supernal hopes and one of the most wonderful promises of religion.

◆

When Darwin originally penned *On the Origin of Species* in 1859, he concluded with this memorable statement:

> There is grandeur in this view of life, with its several powers, having been originally breathed into a few forms or into one;

> and that, whilst this planet has gone cycling on according to the fixed law of gravity, from so simple a beginning endless forms most beautiful and most wonderful have been, and are being, evolved.[4]

In a later edition, however, he inserted three important words into the opening of this final sentence, so that it reads, "There is grandeur in this view of life, with its several powers, having been originally breathed *by the Creator* into a few forms or into one."[5]

Whether Darwin inserted those three words out of genuine belief, or because he felt his largely Christian audience needed to hear it, is anyone's guess.[6] But in my view, at least, there really is grandeur in this view of life—and to fully appreciate it, we need to recognize that we have a divinely ordained purpose. God created us through evolution, and did so in such a way that we are to find life's most profound joys in our family relationships.

Acknowledgments

While it took many years to work out the details of the key ideas described in this book, their origins came to me almost all at once, in a sort of epiphany. I am grateful to the late Clayton M. Christensen for inspiring confidence in me when I expressed these ideas to him shortly following this experience. His early assurance gave me the support, confidence, and resolve I needed to write the book.

I am grateful to Nicholas Christakis for helping me figure out how to get started. Reading his book *Blueprint* rekindled an excitement for my own manuscript after almost a decade of dormancy. My writing has been heavily influenced by his careful research into how evolution has crafted the potential for good in all of us.

I am grateful to many experts for reading over various excerpts or discussing critical topics to improve the technical accuracy of the book, including Roy Baumeister, David Blankenhorn, Björn Brembs, Andrew Cherlin, Simon Conway Morris, Robin Dunbar, Jan Engelmann, Daniel Gilbert, Peter Gollwitzer, Rowell Huesmann, Joshua Knobe, George McGavin, Edward Larson, Christian List, Eddy Nahmias, Ron Numbers, Samir Okasha, Kyle Pruett, Adina Roskies, Robert Sapolsky, Martin Seligman, John Snarey, Elliott Sober, Steve Stearns, Roger

Thomas, Linda Waite, Robert Waldinger, Felix Warneken, Brad Wilcox, and Richard Wrangham. I'm also indebted to many friends and colleagues for reviewing early drafts to improve the readability of the book, including Katherine Bright, Maizy Cloutier, Tammy Cohen, Lia Collings, Patrick Crompton, Ernesto Di Giordano, Mahmood El-Gasim, Terryl Givens, Thomas Griffith, Laura Hernandez, Walter Hill, Rachel Katz, John Morley, Kathy Mouritsen, Eva Murphy, Robert Ostroff, Rose Quiello, Rajiv Radhakrishnan, Greg Rhee, Jeanette Stoneman, Gerrit van Schalkwyk, Ryan Webler, Andrew Wilkinson, and Stuart Williams. I'm grateful to Rose-Lynn Fisher for permission to reprint a photograph from her book.

Joe Sint helped with data analysis. Bailey Fraker did a wonderful job with the illustrations. Toby Lester is a phenomenal editor. Claiborne Hancock, publisher and editor extraordinaire, and the team at Pegasus Books were great to work with. Mark Gottlieb at Trident is an incredibly responsive agent.

For over ten years, I have had the privilege to work at Yale University. Within my little corner of this institution, I've been blessed to be surrounded by incredibly supportive mentors and colleagues.

Most critical was the support of my wife, Sara. Without her love and understanding, none of this would have been possible. My parents, David and Tricia, were also a huge source of support. My five children have been a source of inspiration and I have been richly blessed by their presence in my life. I thank the many teachers I had throughout the years, going back to grade school, for their patience and diligence. And finally, in a book whose chief aim is to reconcile a scientific worldview with a belief in the Divine, I would be remiss not to thank God for support unseen but nonetheless real.

Illustration Credits

Grateful acknowledgment is made to the following for permission to reprint previously published figures: Rose-Lynn Fischer (Figure 3.1) and The Royal Society (Great Britain, Figure 4.1). Grateful acknowledgment is also made to the following for creation of new figures for this book: Bailey Fraker (Figures 2.1 through 2.6) and Kyaw "Joe" Sint (Figure 9.1).

Bibliography

Ackerman, Diane. *A Natural History of Love*. New York: Vintage Books, 1995.

Adams, Noah. "Timeline: Remembering the Scopes Monkey Trial." *National Public Radio*, July 5, 2005. https://www.npr.org/2005/07/05/4723956/timeline-remembering-the-scopes-monkey-trial.

Adams, Tim. "How to Spot a Murderer's Brain." *Guardian*, May 11, 2013. https://www.theguardian.com/science/2013/may/12/how-to-spot-a-murderers-brain.

Aknin, Lara B., Tanya Broesch, J. Kiley Hamlin, and Julia W. Van de Vondervoort. "Prosocial Behavior Leads to Happiness in a Small-Scale Rural Society." *Journal of Experimental Psychology: General* 144, No. 4 (Aug. 2015): 788–795.

Aknin, Lara B., Alice L. Fleerackers, and J. Kiley Hamlin. "Can Third-Party Observers Detect Emotional Rewards of Generous Spending?" *Journal of Positive Psychology* 9, No. 3 (2014): 198–203.

Aknin, Lara B., J. Kiley Hamlin, and Elizabeth W. Dunn. "Giving Leads to Happiness in Young Children." *PLoS ONE* 7, No. 6 (June 2012), https://doi.org/10.1371/journal.pone.0039211.

Alexander, Robert D. *The Biology of Moral Systems*. New York: Adline de Gruyter, 1987.

Allen, Madelene. *Wake of the Invercauld: Shipwrecked in the Sub-Antarctic: A Great-Granddaughter's Pilgrimage*. Dunedin, NZ: Exisle Publishing, 1997.

Anderson, P. W. "More is Different: Broken Symmetry and the Nature of the Hierarchical Structure of Science." *Science* 177, No. 4047 (Aug. 4, 1972): 393–396.

Andersson, Dan I. "Evolution of Antibiotic Resistance." In *Princeton Guide to Evolution*, ed. Jonathan B. Losos, 747–753. Princeton, N.J.: Princeton University Press, 2014.

Antonovics, Kate, and Robert Town. "Are All the Good Men Married? Uncovering the Sources of the Marital Wage Premium." *The American Economic Review* 94, No. 2 (May 2004): 317–321.

Archer, John, Nicola Graham-Kevan, and Michelle Davies. "Testosterone and Aggression: A Reanalysis of Book, Starzyk, and Quinsey's (2001) Study." *Aggression and Violent Behavior* 10 (2005): 241–261.

Argyle, Michael. "Causes and Correlates of Happiness." In *Well-Being: The Foundations of Hedonic Psychology*, eds. Daniel Kahneman, Edward Diener, and Norbert Schwarz, 353–373. New York: Russell Sage Foundation, 1999.

Arnocky, Steven, Tina Piche, Graham Albert, Danielle Oulette, and Pat Barclay. "Altruism Predicts Mating Success in Humans." *British Journal of Psychology* 108, No. 2 (May 2017): 416–435.

Axelrod, Robert. "Effective Choice in the Prisoner's Dilemma." *Journal of Conflict Resolution* 24, No. 1 (March 1, 1980): 3–25.

Bandura, Albert. "Influence of Models' Reinforcement Contingencies on the Acquisition of Imitative Responses." *Journal of Personality and Social Psychology* 1, No. 6 (1965): 589–595.

Bandura, Albert, Dorothea Ross, and Sheila A. Ross. "Imitation of Film-Mediated Aggressive Models." *Journal of Abnormal and Social Psychology* 66, No. 1 (1963): 3–11.

Bandura, Albert, Dorothea Ross, and Sheila A. Ross. "Transmission of Aggression Through Imitation of Aggressive Models." *Journal of Abnormal and Social Psychology* 63, No. 3 (1961): 575–582.

Barash, David P., and Judith Eve Lipton. *The Myth of Monogamy: Fidelity and Infidelity in Animals and People.* New York: W.H. Freeman/Henry Holt and Company, 2002.

Barber, Nigel. "Machiavellianism and Altruism: Effect of Relatedness of Target Persons on Machiavellian and Helping Attitudes." *Psychological Reports* 75 (1994): 403–422.

Bargh, John, Mark Chen, and Lara Burrows. "Automaticity of Social Behavior: Direct Effects of Trait Construct and Stereotype Activation on Action." *Journal of Personality and Social Psychology* 71, No. 2 (1996): 230–244.

Barlow, Connie. "Let There Be Sight! A celebration of convergent evolution." April 2003. http://www.thegreatstory.org/convergence.html.

Barnes, Elizabeth. "Emergence and Fundamentality." *Mind* 121, No. 484 (Oct. 2012): 873–901.

Bateson, Melissa, Daniel Nettle, and Gilbert Roberts. "Cues of Being Watched Enhance Cooperation in a Real-World Setting." *Biology Letters* 2, No. 3 (June 27, 2006): 412–414.

Baumeister, Roy F., E. J. Masicampo, and Kathleen D. Vohs. "Do Conscious Thoughts Cause Behavior?" *Annual Review of Psychology* 62 (2011): 331–361.

Baumeister, Roy F., Kathleen D. Vohs, Jennifer L. Aaker, and Emily N. Garbinsky. "Some Key Differences Between a Happy Life and a Meaningful Life." *Journal of Positive Psychology* 8, No. 6 (2013): 505–516.

Bębenek, Anna, and Izabela Ziuzia-Graczyk. "Fidelity of DNA Replication—A Matter of Proofreading." *Current Genetics* 64 (2018): 985–996.

Belsky, Jay, Lise Youngblade, Michael Rovine, and Brenda Volling. "Patterns of Marital Change and Parent-Child Interaction." *Journal of Marriage and Family* 53, No. 2 (1991): 487–98.

Berger, Lawrence M., Christina Paxson, and Jane Waldfogel. "Mothers, Men, and Child Protective Services Involvement." *Child Maltreatment* 14, No. 3 (Aug. 2009): 263–276

Bhasin, Shalender, Juan P. Brito, Glenn R. Cunningham, Frances J. Hayes, Howard N. Hodis, Alvin M. Matsumoto, Peter J. Snyder, Ronald S. Swedloff, Frederick C. Wu, and Maria A. Yialamas. "Testosterone Therapy in Male with Hypogonadism: An Endocrine Society Clinical Practice Guidelines." *Journal of Clinical Endocrinology and Metabolism* 103, No. 5 (May 2018): 1715–1744.

Biller, Henry B. *Fathers and Families: Paternal Factors in Child Development.* Westport, Conn.: Auburn House, 1993.

Billig, Michael, and Henri Tajfel. "Social Categorization and Similarity in Intergroup Behaviour." *European Journal of Social Psychology* 3, No. 1 (Jan./March 1973): 27–52.

Bjorkland, David F. "The Role of Immaturity in Human Development." *Psychological Bulletin* 122, No. 2 (1997): 153–169.

Blanchflower, David G., and Andrew E. Clark. "Children, Unhappiness and Family Finances." *Journal of Population Economics* 34 (2021): 625–653.

Blanchflower, David G., and Andrew J. Oswald. "International Happiness: A New View on the Measure of Performance." *Academy of Management Perspectives* 25, No. 1 (Feb. 2011): 6–22.

Blankenhorn, David. *Fatherless America: Confronting Our Most Urgent Social Problem.* New York: Harper Perennial, 1995.

Boehm, Christopher. *Moral Origins: The Evolution of Virtue, Altruism, and Shame.* New York: Basic Books, 2012.

Bortolotti, Laura, and Cecilia Costa. "Chemical Communication in the Honey Bee Society." In *Neurobiology of Chemical Communication*, ed. Carla Mucignat-Caretta, 147–210. Boca Raton, Fla.: CRC Press/Taylor & Francis, 2014.

Bowlby, John. *Attachment and Loss: Volume I: Attachment.* New York: Basic Books, 1969.

Bowlby, John. "The Nature of the Child's Tie to His Mother." *The International Journal of Psychoanalysis* 39 (1958): 350–373.

Bowles, Samuel, and D. Posel. "Genetic Relatedness Predicts South African Migrant Workers' Remittances to Their Families." *Nature* 434, No. 7031 (March 17, 2005): 380–383.

Bowles, Samuel. "Conflict: Altruism's Midwife." *Nature* 456, No. (Nov. 20, 2008): 326–327.

Bowles, Samuel. "Did Warfare Among Ancestral Hunter-Gatherers Affect the Evolution of Human Social Behaviors?" *Science* 324, No. 5932 (June 5, 2009): 1293–1298.

Bowles, Samuel. "Group Competition, Reproductive Leveling, and the Evolution of Human Altruism." *Science* 314, No. 5805 (Dec. 8, 2006): 1569–1572.

Brase, Gary L. "Cues of Parental Investment as a Factor in Attractiveness." *Evolution and Human Behavior* 27, No. 2 (March 2006): 145–157.

Brasil-Neto, Joaquim P., Alvaro Pascual-Leone, Josep Valls-Sole, Leonardo G. Cohen, and Mark Hallett. "Focal Transcranial Magnetic Stimulation and Response Bias in a Forced-Choice Task." *Journal of Neurology, Neurosurgery, and Psychiatry* 55 (1992): 964–966.

Brembs, Björn. "The Brain as a Dynamically Active Organ." *Biochemical and Biophysical Research Communications* 564 (July 30, 2021): 55–69.

Brembs, Björn. "Towards a Scientific Concept of Free Will as a Biological Trait: Spontaneous Actions and Decision-Making in Invertebrates." *Proceedings of the Royal Society B* 278 (2011): 930–939.

Brickman, Philip, Dan Coates, and Ronnie Janoff-Bulman. "Lottery Winner and Accident Victims: Is Happiness Relative?" *Journal of Personality and Social Psychology* 36, No. 8 (1978): 917–927.

Briggman, K. L., H.D.I. Abarbanel, and W. B. Kristan Jr. "Optimal Imaging of Neuronal Populations During Decision-Making." *Science* 307, No. 5711 (Feb. 11, 2005): 896–901.

Brink, Susan. *The Fourth Trimester: Understanding, Protecting, and Nurturing an Infant Through the First Three Months.* Berkeley: University of California Press, 2013.

Brooks, David. "The Age of Possibility." *New York Times*, Nov. 16, 2012. nytimes.com/2012/11/16/opinion/brooks-the-age-of-possibility.html.

Brooks, David. *The Second Mountain: The Quest for a Moral Life.* New York: Random House, 2019.

Brown, James R., and W. Ford Doolittle. "*Archae* and the Prokaryote-to-Eukaryote Transition." *Microbiology and Molecular Biology Reviews* 61, No. 4 (1997): 456–502.

Brunner, H. G., M. Nelen, X. O. Breakefield, H. H. Ropers, and B. A. van Oost. "Abnormal Behavior Associated with a Point Mutation in the Structural Gene for Monoamine Oxidase A." *Science* 262, No. 5133 (Oct. 22, 1993): 578–580.

Bryan, William Jennings. *The Prince of Peace*. New York and London: Funk and Wagnalls Company, 1914.

Bryson, Bill. *The Body: A Guide for Occupants*. New York: Doubleday, 2019.

Bullens, Lottie, Jens Forster, Frenk van Harreveld, and Nira Liberman. "Self-Produced Decisional Conflict Due to Incorrect Metacognitions." In *Cognitive Consistency: A Fundamental Principle in Social Cognition*, eds. Bertram Gawronski and Fritz Strack, 285–304. New York: Guilford Press, 2012.

Burch, Rebecca L., and Gordon G. Gallup, Jr. "Abusive Men Are Driven by Paternal Uncertainty." *Evolutionary Behavioral Sciences* 14, No. 2 (April 2020): 197–209.

Burton-Chellew, Maxwell N., and Robin I. M. Dunbar. "Hamilton's Rule Predicts Anticipated Social Support in Humans." *Behavioral Ecology* 26, No. 1 (Jan.–Feb. 2015): 130–137.

Buss, David. *Evolutionary Psychology: The New Science of the Mind*. New York: Routledge, 2015.

Buss, David M. *The Evolution of Desire: Strategies of Human Mating*. New York: Basic Books, 2016.

Buss, David M. "Jealousy, The Necessary Evil." *Los Angeles Times*, Feb. 14, 2007. https://www.latimes.com/archives/la-xpm-2007-feb-14-oe-buss14-story.html.

Buss, David M., Randy J. Larsen, Drew Westen, and Jennifer Semmelroth. "Sex Differences in Jealousy: Evolution, Physiology, and Psychology." *Psychological Science* 3, No. 4 (July 1992): 251–255.

Buss, David M., and David P. Schmitt. "Sexual Strategies Theory: An Evolutionary Perspective on Human Mating." *Psychological Review* 100, No. 2 (1993): 204–232.

Butterfield, Jeremy. "Laws, Causation and Dynamics at Different Levels." *Interface Focus* 2 (Feb. 2012): 101–114.

Calati, Raffaella, Chiara Ferrari, Marie Brittner, Osmano Oasi, Emilie Olié, André F. Carvahlo, and Philippe Courtet. "Suicidal Thoughts and Behaviors and Social Isolation: A Narrative Review of the Literature." *Journal of Affective Disorders* 245 (2019): 653–667.

Campbell, Angus, Philip Converse, and Willard Rodgers. *The Quality of American Life: Perceptions, Evaluations, and Satisfactions*. New York: Russell Sage Foundation, 1976.

Caplan, Bryan. *Selfish Reasons to Have More Kids*. New York: Basic Books, 2011.

Case, Anne, and Angus Deaton. *Deaths of Despair and the Future of Capitalism*. Princeton, N.J.: Princeton University Press, 2020.

Cave, Stephen. "There's No Such Thing as Free Will." *The Atlantic*, June 2016. https://www.theatlantic.com/magazine/archive/2016/06/theres-no-such-thing-as-free-will/480750/.

Centers for Disease Control and Prevention. "WISQARS Fatal Injury Reports, National, Regional and State, 1981–2018." Accessed on Sept. 23, 2021. https://webappa.cdc.gov/sasweb/ncipc/mortrate.html.

Chappell, Bill, and Sasha Ingber. "'This is Not Who We Are,' Colorado Officials Say After Deadly School Shooting." *National Public Radio*, May 8, 2019. https://www.npr.org/2019/05/08/721474989/this-is-not-who-we-are-colorado-officials-say-after-deadly-school-shooting.

Chavez, Nicole, and Marlena Baldacci. "This UNC Charlotte Student Knocked the Shooter Off His Feet and Saved Lives." *CNN*, May 2, 2019. https://www.cnn.com/2019/05/01/us/uncc-victims-shooting-riley-howell/index.html.

Chetty, Raj, John N. Friedman, Nathaniel Hendren, Maggie R. Jones, and Sonya R. Porter. "The Opportunity Atlas: Mapping the Childhood Roots of Social Mobility." NBER Working Paper No. 25147, Oct. 2018, revised Feb. 2020.

Chick, Garry, John W. Loy, and Andrew W. Miracle. "Combative Sport and Warfare: A Reappraisal of the Spillover and Catharsis Hypotheses." *Cross-Cultural Research* 31, No. 3 (1997): 249–267.

Choi, Jung-Kyoo, and Samuel Bowles. "The Coevolution of Parochial Altruism and War." *Science* 318 (Oct. 26, 2007): 636–640.

Christakis, Nicholas. *Blueprint: The Evolutionary Origins of a Good Society.* New York: Little, Brown Spark, 2019.

Christensen, Clayton M., James Allworth, and Karen Dillon. *How Will You Measure Your Life?* New York: HarperCollins, 2012.

Christin, Pascal-Antoine, Daniel M. Weinreich, and Guillaume Besnard. "Causes and Evolutionary Significance of Genetic Convergence." *Trends in Genetics* 26, No. 9 (Sept. 2010): 400–405.

Cigna US Loneliness Index: Survey of 20,000 Americans Examining Behaviors Driving Loneliness in the United States. May 2018. https://www.multivu.com/players/English/8294451-cigna-us-loneliness-survey/docs/IndexReport_1524069371598-173525450.pdf.

Clark, Russell D., and Elain Hatfield. "Gender Differences in Receptivity to Sexual Offers." *Journal of Psychology & Human Sexuality* 2, No. 1 (1989): 39–55.

Cofran, Zachary. "Brain Size Growth in Wild and Captive Chimpanzees (*Pan troglodytes*)," *American Journal of Primatology* 80 (2018): e22876.

Cohen, Adam S. "Harvard's Eugenics Era." *Harvard Magazine*, March–April 2016. https://www.harvardmagazine.com/2016/03/harvards-eugenics-era.

Collins, Francis. *The Language of God: A Scientist Presents Evidence for Belief.* New York: Free Press, 2006.

Conway Morris, Simon. *Life's Solution: Inevitable Humans in a Lonely Universe.* Cambridge, UK: Cambridge University Press, 2004.

Conway Morris, Simon. Map of Life: Convergent Evolution Online. Accessed January 14, 2023. Mapoflife.org.

Conway Morris, Simon. "The Predictability of Evolution: Glimpses into a Post-Darwinian World." *Naturwissenschaften* 96, No. 11 (Nov. 2009): 1313–1337.

Conway Morris, Simon. "Why Are We Here?" Accessed on Sept. 23, 2021. Transcript available at https://www.whyarewehere.tv/people/simon-conway-morris/.

Costello, Mark J., Robert M. May, and Nigel E. Stork. "Can We Name Earth's Species Before They Go Extinct?" *Science*, 339 No. 6118 (Jan. 25, 2013): 413–416.

Cowan, Carolyn Pape, and Philip A. Cowan. "Men's Involvement in Parenthood: Identifying the Antecedents and Understanding the Barriers." In *Men's Transitions to Parenthood: Longitudinal Studies of Early Family Experience*, ed. Phyllis W. Berman and Frank A. Pedersen, 145–174. New York: Lawrence Erlbaum Associates, 1987.

Crow, James F. "Unequal by Nature: A Geneticist's Perspective on Human Differences." *Daedalus: Journal of the American Academy of Arts & Sciences* 131, No. 1 (Winter 2002): 81–88.

Cruver, Brian. *Anatomy of Greed: The Unshredded Truth from an Enron Insider.* New York: Carol & Graf, 2002.

Cunningham, Michael R. "Levites and Brother's Keepers: A Sociobiological Perspective on Prosocial Behavior." *Humboldt Journal of Social Relations* 13, No. 1/2 (1986): 35–67.

Darwin, Charles. Charles Darwin to Joseph Hooker, March 29, 1863. In the *Darwin Correspondence Project.* https://www.darwinproject.ac.uk/letter/DCP-LETT-4065.xml.

Darwin, Charles. *On the Origin of Species by Means of Natural Selection.* New York: Barnes & Noble Classics, 2004 (original version published 1859).

Darwin, Charles. *The Descent of Man, and Selection in Relation to Sex.* London: Penguin Books, 1879.

Davie, Maurice R. *The Evolution of War: A Study of Its Role in Early Societies.* New Haven, Conn.: Yale University Press, 1929.

Dawkins, Richard. *Ancestor's Tale: A Pilgrimage to the Dawn of Evolution.* Boston: Houghton Mifflin, 2004.

Dawkins, Richard. *The Blind Watchmaker: Why the Evidence of Evolution Reveals a Universe without Design.* New York: Norton & Company, 1986.

Dawkins, Richard. *Climbing Mount Improbable.* New York: W.W. Norton & Co., 1996.

Dawkins, Richard. *The Selfish Gene.* Oxford, UK: Oxford University Press, 1989.

Dawson, Deborah A. "Family Structure and Children's Health and Well-Being: Data from the 1988 National Health Interview Survey on Child Health." *Journal of Marriage and the Family* 53, No. 3 (Aug. 1991): 573–584.

de Waal, Frans. *Mama's Last Hug.* New York: W.W. Norton & Co., 2019.

Denno, Deborah. "Courts' Increasing Consideration of Behavioral Genetics Evidence in Criminal Cases: Results of a Longitudinal Study." *Michigan State Law Review* 967 (2011): 980–981.

Denton, Michael, and Craig Marshal. "Protein Folds: Laws of Form Revisited." *Nature* 410, No. 6827 (March 22, 2001): 417.

Denworth, Lydia. *Friendship: The Evolution, Biology, and Extraordinary Power of Life's Fundamental Bond.* New York: W.W. Norton & Co., 2020.

Dobzhansky, Theodore. "Biology, Molecular and Organismic." *American Zoologist* 4, No. 4 (Nov. 1964): 443–452.

Dobzhansky, Theodosius. "Nothing in Biology Makes Sense Except in the Light of Evolution." *American Biology Teacher* 35, No. 3 (March 1973): 125–129.

Doyen, Stéphane, Olivier Klein, Cora-Lise Pichon, and Axel Cleeremans. "Behavioral Priming: It's All in the Mind, but Whose Mind?" *PLOS ONE* 7, No. 1 (Jan. 2012): https://doi.org/10.1371/journal.pone.0029081.

Doyle, Bob. "Jamesian Free Will: The Two-Stage Model of William James." *William James Studies* 5 (2010): 1–28.

Druett, Joan. *Island of the Lost: Shipwrecked at the Edge of the World.* Chapel Hill, N.C.: Algonquin Books of Chapel Hill, 2007.

Dubos, René. *So Human an Animal.* New York: Charles Scribner's Sons, 1968.

Dugatkin, Lee Alan, and Lyudmila Trut. *How to Tame a Fox (and Build a Dog): Visionary Scientists and a Siberian Tale of Jump-Started Evolution.* Chicago: University of Chicago Press, 2017.

Dunfield, Kristen A., and Valerie A. Kuhlmeier. "Intention-mediated Selective Helping in Infancy." *Psychological Science* 21, No. 4 (April 2010): 523–527.

Dunham, Yarrow, Andrew Scott Baron, and Susan Carey. "Consequences of Minimal Group Affiliation in Children." *Child Development* 82, No. 3 (May/June 2011): 793–811.

Dunn, Elizabeth W., Lara B. Aknin, and Michael I. Norton. "Spending Money on Others Promotes Happiness." *Science* 319, No. 5870 (March 21, 2008): 1687–1688.

Dwyer, Jim. "The Doctor Came to Save Lives. The Co-op Board Told Him to Get Lost." *New York Times*, April 3, 2020. https://www.nytimes.com/2020/04/03/nyregion/co-op-board-coronavirus-nyc.html.

Ehrlich, Paul R. *Human Natures: Genes, Cultures, and the Human Prospect.* New York: Penguin, 2000.

ElHage, Alyssa. "How Marriage Makes Men Better Fathers." *Institute for Family Studies*, June 19, 2015. https://ifstudies.org/blog/how-marriage-makes-men-better-fathers.

Ellis, Bruce J., John E. Bates, Kenneth A. Dodge, David M. Fergusson, L. John Horwood, Gregory S. Pettit, and Lianne Woodward. "Does Father Absence Place Daughters at Special Risk for Early Sexual Activity and Teenage Pregnancy?" *Child Development* 74, No. 3 (May-June 2003): 801–821.

Ellis, Lee, Kevin Beaver, and John Wright. *Handbook of Crime Correlates.* San Diego: Academic Press, 2009.

Ellwood, David T. *Poor Support: Poverty in the American Family.* New York: Basic Books, 1988.

Engelmann, Jan M., Esther Herrmann, and Michael Tomasello. "Five-Year Olds, but Not Chimpanzees, Attempt to Manage Their Reputations." *PLoS ONE* 7, No. 10 (Oct. 2010): e48433, https://doi.org/10.1371/journal.pone.0048433.

Engelmann, Jan M., Esther Herrmann, and Michael Tomasello. "The Effects of Being Watched on Resource Acquisition in Chimpanzees and Human Children." *Animal Cognition* 19 (2016): 147–151.

Eyring, Henry. *The Faith of a Scientist.* Salt Lake City: Bookcraft, 1967.

Fehr, Ernst, and Urs Fischbacher. "The Nature of Human Altruism." *Nature* 425 (Oct. 23, 2003): 785–791.

Fernald, Russell D. "Evolving Eyes." *International Journal of Developmental Biology* 48 (2004): 701–705.

Ficks, Courtney A., and Irwin D. Waldman. "Candidate Genes for Aggression and Antisocial Behavior: A Meta-Analysis of Association Studies of the *5HTTLPR* and *MAOA-uVNTR*." *Behavior Genetics* 44 (2014): 427–444.

Fisher, Helen. "'Wilson,' They Said, 'You're All Wet!'" *New York Times*, October 16, 1994. https://www.nytimes.com/1994/10/16/books/wilson-they-said-your-all-wet.html?.

Foster, Lawrence. "Free Love and Feminism: John Humphrey Noyes and the Oneida Community." *Journal of the Early Republic* 1, No. 2 (Summer 1981): 165–183.

Fraley, R. Chris, Claudia C. Brumbaugh, and Michael J. Marks. "The Evolution and Function of Adult Attachment: A Comparative and Phylogenetic Analysis." *Journal of Personality and Social Psychology* 89, No. 5 (2005): 731–746.

Francey, Damien, and Ralph Bergmuller. "Images of Eyes Enhance Investments in a Real-Life Public Good." *PLoS One* 7, No. 5 (May 2012), https://doi.org/10.1371/journal.pone.0037397.

Frankl, Viktor. *Man's Search for Meaning.* New York: Simon and Schuster, 1984.

Franklin, Benjamin. "Old Mistresses Apologue." *Franklin Papers*, June 25, 1745. https://founders.archives.gov/documents/Franklin/01-03-02-0011.

Freud, Sigmund. *Civilization and Its Discontents.* Translated by David McLintock. New York: Penguin, 2004 (original 1930).

Freud, Sigmund. *The Interpretation of Dreams.* Translated by James Strachey. New York: Basic Books, 1955.

Fuda, C.C.S., J. F. Fisher, and S. Mobashery. "β-Lactam Resistance in *Staphylococcus aureus*: The Adaptive Resistance of a Plastic Genome." *Cellular and Molecular Life Sciences* 62 (Sept. 7, 2005): 2617–2633.

Fuentes, Augustin. *Race, Monogamy, and Other Lies They Told You: Busting Myths about Human Nature.* Berkeley: University of California Press, 2012.

Furstenberg, Frank F. "Good Dads—Bad Dads: Two Faces of Fatherhood." In *The Changing American Family and Public Policy,* ed. Andrew Cherlin, 193–218. Washington: Urban Press Institute, 1988.

Furstenberg, Frank F., and Andrew Cherlin. *Divided Families: What Happens to Children When Parents Part.* Cambridge, Mass.: Harvard University Press, 1991.

Garreau, Joel. *Radical Evolution.* New York: Broadway Books, 2005.

Genex Diagnostics. Accessed April 26, 2022. https://www.genexdiagnostics.com/.

Gettler, Lee T., Thomas W. McDade, Alan B. Feranil, and Christopher W. Kuzawa. "Longitudinal Evidence that Fatherhood Decreases Testosterone in Human Males." *Proceedings of the National Academy of Sciences* 108, No. 39 (Sept. 27, 2011): 16194–16199.

Gilbert, Daniel. *Stumbling on Happiness.* New York: Alfred A. Knopf, 2006.

Gilbert, Daniel T., and Jane E. J. Ebert. "Decisions and Revisions: The Affective Forecasting of Changeable Outcomes." *Journal of Personality and Social Psychology* 82, No. 4 (2002): 503–514.

Ginther, Donna K., and Madeline Zavodny. "Is the Male Marriage Premiums Due to Selection? The Effect of Shotgun Weddings on the Return to Marriage." *Journal of Population Economics* 14, No. 2 (June 2001): 313–328.

Glass, Jennifer, Matthew A. Andersson, and Robin W. Simon. "Parenthood and Happiness: Effects of Work-Family Reconciliation Policies in 22 OECD Countries." *American Journal of Sociology* 122, No. 3 (Nov. 2016): 886–929.

Glaze, Lauren E. "Correctional Populations in the United States, 2010." Bureau of Justice Statistics, December 2011. https://bjs.ojp.gov/content/pub/pdf/cpus10.pdf.

Glenn, Norval D., and Charles N. Weaver. "The Contribution of Marital Happiness to Global Happiness." *Journal of Marriage and the Family* 43, No. 1 (1981): 161–168.

Glimcher, Paul W. "Indeterminacy in Brain and Behavior." *Annual Review of Psychology* 56 (2005): 25–56.

Godin, Jean-Guy J., and Heather E. McDonough. "Predator Preference for Brightly Colored Males in the Guppy: A Viability Cost for a Sexually Selected Trait." *Behavioral Ecology* 14, No. 2 (March 2003): 194–200.

Goldstein, Pavel, Irit Weissman-Fogel, Guillaume Dumas, and Simone G. Shamay-Tsoory. "Brain-to-Brain Coupling During Handholding is Associated with Pain Reduction." *Proceedings of the National Academy of Science* 115, No. 11 (March 13, 2018): e2528–e2537.

Gollwitzer, Peter. "Implementation Intentions: Strong Effects of Simple Plans." *American Psychologist* 54, No. 7 (July 1999): 493–503.

Gollwitzer, Peter M., and Paschal Sheeran. "Implementation Intentions and Goal Achievement: A Meta-analysis of Effects and Processes." *Advances in Experimental Social Psychology* 38 (2006): 69–119.

Gould, Stephen J. *Wonderful Life: The Burgess Shale and the Nature of History.* New York: W.W. Norton & Co., 1989.

Gove, Walter, Michael Hughes, and Carolyn Style. "Does Marriage Have Positive Effects on the Psychological Well-Being of the Individual?" *Journal of Health and Social Behavior* 24, No. 2 (1983): 122–131.

Gray, Peter. *Psychology.* New York: Worth Publishers, 2002.

Grienzi, Greg. "A Remarkable Partnership Sets High-Schoolers on New Path." *The JHU Gazette*, June 11, 2007. https://pages.jh.edu/~gazette/2007/11jun07/11mentor.html.

Grobstein, Paul. "Variability in Brain Function and Behavior." In *The Encyclopedia of Human Behavior, Volume 4*, ed. V. S. Ramachandran, 447–458. Cambridge, Mass.: Academic Press, 1994.

Grover, Shawn, and John F. Helliwell. "How's Life at Home? New Evidence on Marriage and the Set Point for Happiness." *Journal of Happiness Studies* 20 (2019): 373–390.

Gutmann, David. "Parenthood: A Key to Comparative Study of the Life Cycle." In *Life-Span Developmental Psychology*, ed. by Nancy Datan and Leon H. Ginsberg, 167–184. Cambridge, Mass.: Academic Press, 1975.

Hagerty, Barbara Bradley. "Can Your Genes Make You Murder?" *National Public Radio*, July 1, 2010. https://www.npr.org/templates/story/story.php?storyId=128043329.

Haggard, Patrick, and Martin Eimer. "On The Relation Between Brain Potentials and the Awareness of Voluntary Movements." *Experimental Brain Research* 126 (1999): 128–136.

Haidt, Jonathan. *The Happiness Hypothesis: Finding Modern Truth in Ancient Wisdom*. New York: Basic Books, 2006.

Hamilton, W. D. "Innate Social Aptitudes of Man: An Approach from Evolutionary Genetics." In *ASA Studies 4: Biosocial Anthropology*, ed. Robin Fox, 133–135. London: Malaby Press, 1975.

Handelsman, David J., Angelica L. Hirschberg, and Stephane Bermon. "Circulating Testosterone as the Hormonal Basis of Sex Differences in Athletic Performance." *Endocrine Reviews* 39, No. 5 (Oct. 2018): 803–829.

Hare, Brian, and Michael Tomasello. "Human-like Social Skills in Dogs?" *Trends in Cognitive Science* 9, No. 9 (Sept. 2005): 439–444.

Hare, Brian, Victoria Wobber, and Richard Wrangham. "The Self-Domestication Hypothesis: Evolution of Bonobo Psychology is Due to Selection Against Aggression." *Animal Behaviour* 83, No. 3 (2012): 573–585.

Harknett, Kristen, and Arielle Kuperberg. "Education, Labor Markets, and the Retreat from Marriage." *Social Forces* 90, No. 1 (Sept. 1, 2011): 41–63.

Harlow, Harry. "The Nature of Love." *American Psychologist* 13 (1958): 673–685.

Harris, Sam. *Free Will*. New York: Free Press, 2012.

Harris, Sam. "The Strange Case of Francis Collins." *Sam Harris Blog*, Aug. 6, 2009. https://samharris.org/the-strange-case-of-francis-collins/.

Hawkes, Kristen. "Grandmothers and the Evolution of Human Longevity." *American Journal of Human Biology* 15 (2003): 380–400.

Hawkley, Louise C., and John T. Cacioppo. "Loneliness Matters: A Theoretical and Empirical Review of Consequences and Mechanisms." *Annals of Behavioral Medicine* 40, No. 2 (July 22, 2010): 218–227.

Hazan, Cindy, and Phillip Shaver. "Romantic Love Conceptualized as an Attachment Process." *Journal of Personality and Social Psychology* 52, No. 3 (1987): 511–524.

Heider, Karl G. *Grand Valley Dani: Peaceful Warriors*. Belmont, Calif.: Wadsworth, 1997.

Heisenberg, Martin. "Is Free Will an Illusion?" *Nature* 459 (May 14, 2009): 164–165.

Henrich, Joseph, Robert Boyd, and Peter J. Richerson. "The Puzzle of Monogamous Marriage." *Philosophical Transactions of the Royal Society B* 367 (2012): 657–669.

Herrera, Sebastian. "With Free Haircuts, Houstonian Changes One Disadvantaged Life at a Time." *Houston Chronicle*, Nov. 22, 2017. https://www.houstonchronicle.com/about/houston-gives/article/With-free-haircuts-Houstonian-changes-one-12374650.php.

Hill, Kim R., Robert S. Walker, Miran Božičević, James Eder, Thomas Headland, Barry Hewlett, A. Magdalena Hurtado, Frank Marlow, Polly Wiessner, and Brian Wood. "Co-Residence Patterns in Hunter-Gatherer Societies Show Unique Human Social Structure." *Science* 331, No. 6022 (March 11, 2011): 1286–1289.

Hinckley, Gordon B. "Great Shall Be the Peace of Thy Children." *General Conference Address of The Church of Jesus Christ of Latter-day Saints*, October 7, 2000.

Holt-Lunstad, Julianna, Timothy B. Smith, Mark Baker, Tyler Harris, and David Stephenson. "Loneliness and Social Isolation as Risk Factors for Mortality: A Meta-Analytic Review." *Perspectives on Psychological Science* 10, No. 2 (March 11, 2015): 227–237.

Hooven, Carole. *T: The Story of Testosterone, the Hormone that Divides and Dominates Us.* New York: Henry Holt and Company, 2021.

Horns, Joshua, Rebekah Jung, and David Carrier. "*In vitro* Strain in Human Metacarpal Bones During Striking: Testing the Pugilism Hypothesis of Hominin Hand Evolution." *Journal of Experimental Biology* 218, No. 20 (2015): 3215–3221.

Horwitz, Allan V., Helene Raskin White, and Sandra Howell-White. "Becoming Married and Mental Health: A Longitudinal Study of a Cohort of Young Adults." *Journal of Marriage and the Family* 58 (Nov. 1996): 895–907.

Hoyle, Fred. *The Intelligent Universe.* New York: Holt, Rinehart, and Winston, 1983.

Hoyle, Fred. "The Universe: Past and Present Reflections." *Engineering and Science* (Nov. 1981): 8–12.

Hoyt, Alia. "Do People and Bananas Really Share 50 Percent of the Same DNA?" *HowStuffWorks*, April 1, 2021. https://science.howstuffworks.com/life/genetic/people-bananas-share-dna.htm.

Hrdy, Sarah. *Mothers and Others: The Evolutionary Origins of Mutual Understanding.* Cambridge, Mass.: Belknap Press, 2009.

Huesmann, L. Rowell. "The Contagion of Violence." In *The Cambridge Handbook of Violent and Aggressive Behavior*, eds. Alexander T. Vazsonyi, Daniel J. Flannery, and Matt DeLisi, 527–556. Cambridge, UK: Cambridge University Press, 2018.

Hughes, Philip M., Tabitha L. Ostrout, Mónica Pèrez, and Kathleen C. Thomas. "Adverse Childhood Experiences Across Birth Generation and LGBTQ+ Identity, Behavioral Risk Factor Surveillance System, 2019." *American Journal of Public Health* 112, No. 4 (2022): 662–670.

Hughes, Trevor, and Rebecca Plevin. "Her 'Final Good Deed': Woman Hailed as Hero After Taking Bullets to Protect Rabbi During Synagogue Shooting." *USA Today*, April 28, 2019. https://www.usatoday.com/story/news/2019/04/28/lori-gilbert-kaye-woman-protected-rabbi-synagogue-shooting/3608418002/.

Huxley, Thomas Henry. Thomas Henry Huxley to Charles Darwin, November 23, 1859. In the *Darwin Correspondence Project*. https://www.darwinproject.ac.uk/letter/DCP-LETT-2544.xml.

International Chicken Genome Sequencing Consortium. "Sequence and Comparative Analysis of the Chicken Genome Provide Unique Perspectives on Vertebrate Evolution." *Nature* 432, No. 7018 (Dec. 9, 2004): 695–777.

Jankowiak, William R., and Edward F. Fischer. "A Cross-Cultural Perspective on Romantic Love." *Ethnology* 31, No. 2 (April 1992): 149–155.

Jordan, Matthew R., Jillian J. Jordan, and David G. Rand. "No Unique Effect of Intergroup Competition on Cooperation: Noncompetitive Thresholds Are as Effective as Competitions Between Groups for Increasing Human Cooperative Behavior." *Evolution and Human Behavior* 38 (2017): 102–108.

Kahneman, Daniel. *Thinking, Fast and Slow*. New York: Farrar, Straus and Giroux, 2011.

Kalmijn, Matthijs. "The Effects of Divorce on Men's Employment and Social Security Histories." *European Journal of Population* 21 (2005): 347–366.

Kaplan, Hillard, Kim Hill, Jane Lancaster, and A. Magdalena Hurtado. "A Theory of Human Life History Evolution: Diet, Intelligence, and Longevity." *Evolutionary Anthropology* 9, No. 4 (2000): 156–185.

Keil, Margaret F., Adela Leahu, Megan Rescigno, Jennifer Myles, and Constantine A. Stratakis. "Family Environment and Development in Children Adopted from Institutionalized Care." *Pediatric Research* 91 (2022): 1562–1570.

Keller, I., and H. Heckhausen. "Readiness Potentials Preceding Spontaneous Motor Acts: Voluntary vs. Involuntary Control." *Electroencephalography and Clinical Neurophysiology* 76 (1990): 351–361.

Klaw, Spencer. *Without Sin: The Life and Death of the Oneida Community*. New York: Penguin, 1994.

Konnikova, Maria. "Revisiting Robbers Cave: The Easy Spontaneity of Intergroup Conflict." *Scientific American*, September 5, 2012. https://blogs.scientificamerican.com/literally-psyched/revisiting-the-robbers-cave-the-easy-spontaneity-of-intergroup-conflict/.

Korenman, Sanders, and David Neumark. "Does Marriage Really Make Men More Productive?" *Journal of Human Resources* 26, No. 2 (Spring 1991): 282–307.

Kosslyn, Stephen M., and Samuel T. Moulton. "Mental Imagery and Implicit Memory." In *The Handbook of Imagination and Mental Simulation*, eds. Keith D. Markman, William M. P. Klein, and Julie A. Suhr, 35–51. New York: Psychology Press, 2009.

Kosztolányi, A., Z. Barta, C. Küpper, and T. Székely. "Persistence of an Extreme Male-Biased Adult Sex Ratio in a Natural Population of Polyandrous Bird." *Journal of Evolutionary Biology* 24, No. 8 (Aug. 2011): 1842–1846.

Kushnir, Tamar, Alison Gopnik, Nadia Chernyak, Elizabeth Seiver, and Henry M. Wellman. "Developing Intuitions About Free Will Between Ages Four and Six." *Cognition* 138 (2015): 79–101.

Lancaster, Jane B., and Chet S. Lancaster. "The Watershed: Change in Parental-Investment and Family-Formation Strategies in the Course of Human Evolution." In *Parenting Across the Lifespan*, ed. Jane B. Lancaster, 187–205. New York: Aldine Publishing Co., 2010.

Larson, Edward J. *Summer for the Gods: The Scopes Trial and America's Continuing Debate Over Science and Religion.* New York: Basic Books, 1997.

Leakey, Richard, and Roger Lewin. *Origins: What New Discoveries Reveal About the Emergence of Our Species and its Possible Future.* New York: E.P. Dutton, 1977.

Lee, Kristen Schultz, and Hiroshi Ono. "Marriage, Cohabitation, and Happiness: A Cross-National Analysis of 27 Countries." *Journal of Marriage and Family* 74 (Oct. 2012): 953–972.

Lennon, Jay T., and Kenneth J. Locey. "More Support for Earth's Massive Microbiome." *Biology Direct* 15, No. 5 (March 4, 2020): https://doi.org/10.1186/s13062-020-00261-8.

Lerman, Robert I., and W. Bradford Wilcox. *For Richer, For Poorer: How Family Structures Economic Success in America.* Washington: American Enterprise Institute, 2014.

Levin, Mark, Amy Prosser, David Evans, and Stephen Reicher. "Identity and Emergency Intervention: How Social Group Membership and Inclusiveness of Group Boundaries Shape Helping Behavior." *Personality and Social Psychology Bulletin* 31, No. 4 (April 1, 2005): 443–453.

Libet, Benjamin, Curtis A. Gleason, Elwood W. Wright, and Dennis K. Pearl. "Time of Conscious Intention to Act in Relation to Onset of Cerebral Activity (Readiness Potential): The Unconscious Initiation of a Freely Voluntary Act." *Brain* 106 (1983): 623–642.

Lichter, Daniel T., Diane K. McLaughlin, and David C. Ribar. "Economic Restructuring and the Retreat from Marriage." *Social Science Research* 31, No. 2 (June 2002): 230–256.

Lippa, Richard A. "Sex Differences in Sex Drive, Sociosexuality, and Height Across 53 Nations: Testing Evolutionary and Social Structural Theories." *Archives of Sexual Behavior* 38 (2009): 631–651.

List, Christian. *Why Free Will is Real.* Cambridge, Mass.: Harvard University Press, 2019.

Lizardi, Dana, Ronald G. Thompson, Katherine Keyes, and Deborah Hasin. "Parental Divorce, Parental Depression, and Gender Differences in Adult Offspring Suicide Attempt." *Journal of Nervous and Mental Disease* 197, No. 12 (Dec. 2009): 899–904.

Losos, Jonathan. "What is Evolution?" In *The Princeton Guide to Evolution*, ed. Jonathan Losos. Princeton, N.J.: Princeton University Press, 2014, 3–9.

Loyau, Adeline, Michel Saint Jalme, Cécile Cagniant, and Gabriele Sorci. "Multiple Sexual Advertisements Honestly Reflect Health Status in Peacocks (*Pavo cristatus*)." *Behavioral Ecology and Sociobiology* 58 (May 3, 2005): 552–557.

Lukas, D., and T. H. Clutton-Brock. "Evolution of Social Monogamy." *Science* 341, No. 6145 (Aug. 2, 2013): 526–530.

Madsen, Elaine A., Richard J. Tunney, George Fieldman, Henry C. Plotkin, Robin I. M. Dunbar, Jean-Marie Richardson, and David McFarland. "Kinship and Altruism: A Cross-Cultural Experimental Study." *British Journal of Psychology* 98, No. 2 (May 2007): 339–359.

Maier, Uwe-G, Stefan Zauner, Christian Woehle, Kathrin Bolte, Franziska Hempel, John F. Allen, and William F. Martin. "Massively Convergent Evolution for Ribosomal Protein Gene Content in Plastic and Mitochondrial Genomes." *Genome Biology and Evolution* 5, No. 12 (2013): 2318–2329.

Malthus, Thomas Robert. *An Essay on the Principle of Population.* 1798.

Marean, Curtis. "An Evolutionary Anthropological Perspective on Modern Human Origins." *Annual Review of Anthropology* 44 (Oct. 2015): 533–556.

Margolin, Leslie. "Child Abuse by Mothers' Boyfriends: Why the Overrepresentation?" *Child Abuse & Neglect* 16 (1992): 541–551.

Marks, Nadine F., and James D. Lambert. "Marital Status Continuity and Change Among Young and Midlife Adults: Longitudinal Effects on Psychological Well-being." *Journal of Family Issues* 19, No. 6 (Nov. 1998): 652–686.

Mastekaasa, Arne. "Marital Status, Distress, and Well-Being: An International Comparison." *Journal of Comparative Family Studies* 25, No. 2 (Summer 1994): 183–205.

Maye, Alexander, Chih-hao Hsieh, George Sugihara, and Björn Brembs. "Order in Spontaneous Behavior." *PLoS ONE* 5 (May 16, 2007).

McDermott, Rose, Dustin Tingley, Jonathan Cowden, Giovanni Frazzetto, and Dominic D. P. Johnson. "Monoamine Oxidase A Gene (MAOA) Predicts

Behavioral Aggression Following Provocation." *Proceedings of the National Academy of Sciences* 106, No. 7 (Feb. 17, 2009): 2118–2123.

McElroy, Damien. "Amish Killer's Widow Thanks Families of Victims for Forgiveness." *The Daily Telegraph*, October 16, 2006. https://www.telegraph.co.uk/news/worldnews/1531570/Amish-killers-widow-thanks-families-of-victims-for-forgiveness.html.

McGhee, George. "Convergent Evolution: A Periodic Table of Life?" In *The Deep Structure of Biology*, ed. by Simon Conway Morris, 17–31. West Conshohocken, Penn.: Templeton Press, 2008.

McIntyre, Matthew, Steven W. Gangestad, Peter B. Gray, Judith Flynn Chapman, Terence C. Burnham, Mary T. O'Rourke, and Randy Thornhill. "Romantic Involvement Often Reduces Men's Testosterone Levels—But Not Always: The Moderating Role of Extrapair Sexual Interest." *Journal of Personality and Social Psychology* 91, No. 4 (Oct. 2006): 642–651.

McLanahan, Sara, and Gary Sandefur. *Growing Up With a Single Parent: What Hurts, What Helps*. Cambridge, Mass.: Harvard University Press, 1994.

Mele, Alfred. *Free: Why Science Hasn't Disproved Free Will.* Oxford, UK: Oxford University Press, 2014.

Meyer, Daniel R., Marcia J. Carlson, and Md Moshi Ul Alam. "Increases in Shared Custody after Divorce in the United States." *Demographic Research* 46, No. 38 (2022): 1137–1162.

Miller, Kenneth. *The Human Instinct: How We Evolved to Have Reason, Consciousness, and Free Will.* New York: Simon & Schuster, 2018.

Miller, Stanley L. "A Production of Amino Acids Under Possible Primitive Earth Conditions." *Science* 117, No. 3046 (May 15, 1953): 528–529.

Milne, Sarah, Sheina Orbell, and Paschal Sheeran. "Combining Motivational and Volitional Interventions to Promote Exercise Participation: Protection Motivation Theory and Implementation Intentions." *British Journal of Health Psychology* 7 (2002): 163–184.

Mok, Pearl L. H., Aske Astrup, Matthew J. Carr, Sussie Antonsen, Roger T. Webb, and Carsten B. Pedersen. "Experience of Child-Parent Separation and Later Risk of Violent Criminality." *American Journal of Preventive Medicine* 55, No. 2 (Aug. 2018): 178–186.

Muir, William M., and David Sloan Wilson. "When the Strong Outbreed the Weak: An Interview with William Muir." *This View of Life*, July 11, 2016. https://thisviewoflife.com/when-the-strong-outbreed-the-weak-an-interview-with-william-muir/.

Murgraff, Vered, David White, and Keith Phillips. "Moderating Binge Drinking: It Is Possible to Change Behaviour If You Plan It in Advance." *Alcohol and Alcoholism* 31, No. 6 (1996): 577–582.

Myers, David G. "The Funds, Friends, and Faith of Happy People." *American Psychologist* 55, No. 1 (2000): 56–67.

Nahmias, Eddy. "Is Free Will an Illusion? Confronting Challenges from the Modern Mind Sciences." In *Moral Psychology, Vol. 4: Freedom and Responsibility*, ed. Walter Sinnott-Armstrong, 1–57. Cambridge, Mass.: MIT Press, 2014.

Nahmias, Eddy. "Scientific Challenges to Free Will." In *A Companion to the Philosophy of Action*, eds. C. Sandis and T. O'Connor, 345–356. Hoboken, N.J.: Wiley-Blackwell, 2010.

Nahmias, Eddy. "Why We Have Free Will." *Scientific American*, Jan. 1, 2015, 76–80.

National Center for Education Statistics. "Characteristics of Children's Families." Last updated in May 2022. https://nces.ed.gov/programs/coe/indicator/cce.

National Commission on America's Urban Families. *Families First: Report of the National Commission on America's Urban Families*. Washington: U.S. Government Printing Office, 1993.

National Human Genome Research Institute. "Why Mouse Matters." Last updated July 23, 2010. https://www.genome.gov/10001345/importance-of-mouse-genome.

Nibley, Hugh. *Old Testament and Related Studies*. Salt Lake City: Deseret Book Company, 1986.

Nickle, David C., and Leda M. Goncharoff. "Human Fist Evolution: A Critique." *Journal of Experimental Biology* 216, No. 12 (June 2013): 2359–2360.

Nicholi, Armand M. "The Impact of Family Dissolution on the Emotional Health of Children and Adolescents." In *When Families Fail . . . The Social Costs*, ed. Bryce J Christensen, 27–41. Lanham, Md.: University Press of America, 1991.

Nichols, Shaun. "Experimental Philosophy and the Problem of Free Will." *Science* 331, No. 6023 (March 18, 2011): 1401–1403.

Noyes, John H. *Confessions of John H. Noyes. Part I. Confession of Religious Experience: Including a History of Modern Perfectionism*. Oneida, N.Y.: Leonard & Company Printers, 1849.

O'Brien, Kathleen. "David Popenoe: The Marriage Dude." *Inside New Jersey*, Feb. 11, 2007. https://www.nj.com/iamnj/2007/02/david_popenoe.html.

Okasha, Samir. *Philosophy of Biology: A Very Short Introduction*. Oxford, UK: Oxford University Press, 2019.

Oliver, William J., Lawrence R. Kuhns, and Elaine S. Pomeranz. "Family Structure and Child Abuse." *Clinical Pediatrics* 45, No. 2 (March 2006): 111–118.

Orbell, Sheina, Sarah Hodgkins, and Paschal Sheeran. "Implementation Intentions and the Theory of Planned Behavior." *Personality and Social Psychology Bulletin* 23, No. 9 (1997): 945–954.

Paine, Amy L., Oliver Perra, Rebecca Anthony, and Katherine H. Shelton. "Charting the Trajectories of Adopted Children's Emotional and Behavioral Problems: The Impact of Early Adversity and Postadoptive Parental Warmth." *Development and Psychopathology* 33 (2021): 922–936.

Park, Travis, Erich M. G. Fitzgerald, and Alistair R. Evans. "Ultrasonic Hearing and Echolocation in the Earliest Toothed Whales." *Biology Letters* 12 (April 1, 2016), 20160060, doi.org/10.1098/rsbl.2016.0060.

Perales, Francisco, Sarah E. Johnson, Janeen Baxter, David Lawrence and Stephen R. Zubrick. "Family Structure and Childhood Mental Disorders: New Findings from Australia." *Social Psychiatry and Psychiatric Epidemiology* 52 (2017): 423–433.

Pew Research Center. "Americans, Politics and Science Issues." July 1, 2015. https://www.pewresearch.org/science/2015/07/01/chapter-4-evolution-and-perceptions-of-scientific-consensus/.

Pew Research Center. "Few Americans Blame God or Say Faith Has Been Shaken Amid Pandemic, Other Tragedies." Nov. 23, 2021. https://www.pewresearch.org/religion/2021/11/23/few-americans-blame-god-or-say-faith-has-been-shaken-amid-pandemic-other-tragedies/.

Pew Research Center. "Where Americans Find Meaning in Life." Nov. 20, 2018. https://www.pewresearch.org/religion/2018/11/20/where-americans-find-meaning-in-life/.

Pinker, Steven. *How the Mind Works.* New York and London: W.W. Norton & Company, 1997.

Pinker, Steven, and Paul Bloom. "Natural Language and Natural Selection." *Behavioral and Brain Sciences* 13 (1990): 707–727.

Pinsker, Joe. "It Isn't the Kids. It's the Cost of Raising Them." *The Atlantic*, Feb. 27, 2019. https://www.theatlantic.com/family/archive/2019/02/cost-raising-kids-parents-happiness/583699/.

Ponomarenko, Elena A., Ekaterina V. Poverennaya, Ekaterina V. Ilgisonis, Mikhail A. Pyatnitskiy, Arthur T. Kopylov, Victor G. Zgoda, Andrey V. Lisitsa, and Alexander I. Archakov. "The Size of the Human Proteome: The Width and Depth." *International Journal of Analytical Chemistry* (2016): Article ID 7436849, https://doi.org/10.1155/2016/7436849.

Pontius, Joan U., James C. Mullikin, Douglas R. Smith, Agencourt Sequencing Team, Kerstin Lindblad-Toh, et al. "Initial Sequence and Comparative Analysis of the Cat Genomes." *Genome Research* 17, No. 11 (Nov. 2007): 1675–1689.

Popenoe, David. *Life Without Father: Compelling New Evidence that Fatherhood and Marriage Are Indispensable for the Good of Children and Society.* New York: The Free Press, 1996.

Popenoe, David. *War Over the Family.* New Brunswick, N.J.: Transaction Publishers, 2005.

Prather, Dirk C. "Prompted Mental Practice as a Flight Simulator." *Journal of Applied Psychology* 57, No. 3 (1973): 353–355.

Prüfer, Kay, Kasper Munch, Ines Hellmann, Keiko Akagi, James R. Miller, Brian Walenz, Sergey Koren, et al. "The Bonobo Genome Compared with the Chimpanzee and Human Genomes." *Nature* 486, No. 7404 (June 28, 2012): 527–531.

Pruitt, Jonathan N., and Charles J. Goodnight. "Site-Specific Group Selection Drives Locally Adapted Group Compositions." *Nature* 514, No. 7522 (Oct. 16, 2014): 359–362.

Putnam, Robert. *Bowling Alone: The Collapse and Revival of American Community.* New York: Simon & Schuster, 2000.

Quammen, David. *The Tangled Tree: A Radical New History of Life.* New York: Simon & Schuster, 2018.

Radhakrishna, Aruna, Ingrid E. Bou-Saada, Wanda M. Hunter, Diane J. Catellier, and Jonathan B. Kotch. "Are Father Surrogates a Risk Factor for Child Maltreatment?" *Child Maltreatment* 6, No. 4 (Nov. 2001): 281–289.

Raff, Rudolph, Charles R. Marshall, and James M. Turbeville. "Using DNA Sequences to Unravel the Cambrian Radiation of the Animal Phyla." *Annual Review of Ecology and Systematics* 25 (1994): 351–375.

Raine, Adrian, J. Reid Meloy, Susan Bihrle, Jackie Stoddard, Lori LaCasse, and Monte S. Buchsbaum. "Reduced Prefrontal and Increased Subcortical Brain Functioning Assessed Using Positron Emission Tomography in Predatory and Affective Murderers." *Behavioral Sciences & The Law* 16, No. 3 (Summer 1998): 319–332.

Raine, Adrian, Monte S. Buchsbaum, and Lori LaCasse. "Brain Abnormalities in Murderers Indicated by Positron Emission Tomography." *Biological Psychiatry* 42, No. 6 (Sept. 15, 1997): 495–508.

Raup, David M. "Geometric Analysis of Shell Coiling: General Problems." *Journal of Paleontology* 40, No. 5 (Sept. 1966): 1178–1190.

Reis, Harry T., and Shelly L. Gable, "Toward a Positive Psychology of Relationships." In *Flourishing: Positive Psychology and the Life Well-Lived*, eds. Corey L. M. Keyes and Jonathan Haidt, 129–159. Washington: American Psychological Association, 2003.

Renwick, Trudi J. *Poverty and Single Parent Families.* New York: Garland Publishing, 1998.

Roney, James R., Katherine N. Hanson, Kristina M. Durante, and Dario Maestripieri. "Reading Men's Faces: Women's Mate Attractiveness Judgments

Track Men's Testosterone and Interest in Infants." *Proceedings of the Royal Society B* 273 (2006): 2169–2175.

Sampson, Robert J., John H. Laub, and Christopher Wimer. "Does Marriage Reduce Crime? A Counterfactual Approach to Within-Individual Causal Effects." *Criminology* 44, No. 3 (2006): 465–508.

Sandeen, Ernest R. "John Humphrey Noyes as the New Adam." *Church History* 40, No. 1 (1971): 82–90.

Sanders, Charles W., Mark Sadoski, Richard M. Wasserman, Robert Wiprud, Mark English, and Rachel Bramson. "Comparing the Effects of Physical Practice and Mental Imagery Rehearsal on Learning Basic Venipucture by Medical Students." *Imagination, Cognition, and Personality* 27, No. 2 (2007): 117–127.

Sanders, Robert. "UC Berkeley Nobelist Charles Townes to Receive $1.5 Million Templeton Prize for Linking Religion and Science." *UC Berkeley News*, March 9, 2005, https://www.berkeley.edu/news/media/releases/2005/03/09_templetonaward.shtml.

Sandomir, Richard. "Victoria Ruvolo, Who Forgave Her Attacker, Is Dead at 59." *New York Times*, March 28, 2019. https://www.nytimes.com/2019/03/28/obituaries/victoria-ruvolo-dead.html.

Sapolsky, Robert. *Behave: The Biology of Humans at Our Best and Worst.* New York: Penguin, 2017.

Scheele, Dirk, Andrea Wille, Keith M. Kendrick, Birgit Stoffel-Wagner, Benjamin Becker, Onur Güntürkün, Wolfgang Maier, and René Hurlemann. "Oxytocin Enhances Brain Reward System Responses in Men Viewing the Face of Their Female Partner." *Proceedings of the National Academy of Sciences* 110, No. 50 (Dec. 2013): 20308–20313.

Schmitt-Kopplin, Philippe, Zelimir Gabelica, Régis D. Gougeon, Agnes Fekete, Basem Kanawati, Mourad Harir, Istvan Gebefuegi, Gerhard Eckel, and Norbert Hertkorn. "High Molecular Diversity of Extraterrestrial Organic Matter in Murchison Meteorite Revealed 40 Years After Its Fall." *Proceedings of the National Academy of Sciences* 107, No. 7 (Feb. 2010): 2763–2768.

Schurger, Aaron, Pengbo "Ben" Hu, Joanna Pak, and Adina L. Roskies. "What is the Readiness Potential?" *Trends in Cognitive Science* 25, No. 7 (July 2021): 558–570.

Schwartz, Barry. *The Paradox of Choice.* New York: HarperCollins, 2004.

Seligman, Martin E. P. *Flourish: A Visionary New Understanding of Happiness and Well-Being.* New York: Atria, 2011.

Senior, Jennifer. *All Joy and No Fun: The Paradox of Modern Parenthood.* New York: HarperCollins, 2014.

Seyfarth, Robert M., Dorothy L. Cheney, and Peter Marler. "Vervet Monkey Alarm Calls: Semantic Communication in a Free-Ranging Primate." *Animal Behaviour* 28, No. 4 (1980): 1070–1094.

Shapiro, Joseph P., and Joannie M. Schrof. "Honor Thy Children." *US News and World Report* 118, No. 8 (Feb. 27, 1995): 38–46.

Shenk, Joshua Wolf. "What Makes Us Happy." *The Atlantic*, June 2009. https://www.theatlantic.com/magazine/archive/2009/06/what-makes-us-happy/307439/.

Sherif, Muzafer, O. J. Harvey, B. Jack White, William R. Hood, and Carolyn W. Sherif. *Intergroup Conflict and Cooperation: The Robbers Cave Experiment.* Norman, Okla.: Institute of Good Relations, University of Oklahoma, 1961.

Sherman, Robert Trattner, and Craig A. Anderson. "Decreasing Premature Termination from Psychotherapy." *Journal of Social and Clinical Psychology* 5, No. 3 (1987): 298–312.

Silk, Joan B., Sarah F. Brosnan, Jennifer Vonk, Joseph Henrich, Daniel J. Povinelli, Amanda S. Richardson, Susan B. Lambeth, Jenny Mascaro, and Steven J. Schapiro. "Chimpanzees Are Indifferent to the Welfare of Unrelated Group Members." *Nature* 437, No. 7063 (2005): 1357–1359.

Silver, Laura, Patrick Van Kessel, Christine Huang, Laura Clancy, and Sneha Gubbala. "What Makes Life Meaningful? Views from 17 Advanced Economies." *Pew Research Center*, Nov. 18, 2021, available at https://www.pewresearch.org/global/2021/11/18/what-makes-life-meaningful-views-from-17-advanced-economies/.

Sipes, Richard G. "War, Sports and Aggression: An Empirical Test of Two Rival Theories." *American Anthropologist* 75 (1973): 64–84.

Slutkin, Gary, Charles Ransford, and R. Brent Decker. "Cure Violence: Treating Violence As a Contagious Disease." In *Envisioning Criminology*, eds. Michael D. Maltz and Stephen K. Rice, 45–56. Switzerland: Springer, 2015.

Smith, Adam. *An Inquiry into the Nature and Causes of the Wealth of Nations.* London, 1776.

Snarey, John. *How Fathers Care for the Next Generation: A Four-Decade Study.* Cambridge, Mass.: Harvard University Press, 1993.

Sober, Elliott, and David Sloan Wilson. *Unto Others: The Evolution and Psychology of Unselfish Behavior.* Cambridge, Mass.: Harvard University Press, 1998.

Soma, K. K. "Testosterone and Aggression: Berthold, Birds and Beyond." *Journal of Neuroendocrinology* 18, No. 7 (July 2006): 543–544.

Soon, Chun Siong, Marcel Brass, Hans-Jochen Heinze, and John-Dylans Haynes. "Unconscious Determinants of Free Decisions in the Human Brain." *Nature Neuroscience* 11 (2008): 543–545.

Spencer, Herbert. *Principles of Biology*. New York and London: Williams & Norgate, 1864.

Spitz, René A. "Hospitalism: An Inquiry into the Genesis of Psychiatric Conditions in Early Childhood." *The Psychoanalytic Study of the Child* 1, No. 1 (1945): 53–74.

Spitz, René A. "Hospitalism: A Follow-Up Report on Investigation Described in Volume I, 1945." *The Psychoanalytic Study of the Child* 2, No. 1 (1946): 113–117.

Stack, Steven, and J. Ross Eshleman. "Marital Status and Happiness: A 17-Nation Study." *Journal of Marriage and Family* 60, No. 2 (May 1998): 527–536.

Stearns, Stephen C. "Natural Selection, Adaptation, and Fitness: Overview." In *Princeton Guide to Evolution*, ed. Jonathan B. Losos, 193–199. Princeton, N.J.: Princeton University Press, 2014.

Stevenson, Robert Louis. *The Strange Case of Dr. Jekyll and Mr. Hyde*. London: Longmans, Green, and Co., 1886.

Stiny, Andy. "'Warrior Gene' Defense Mounted in Santa Fe Murder Case." *Albuquerque Journal*, Oct. 24, 2014. https://www.abqjournal.com/485234/warrior-gene-defense-mounted-in-santa-fe-murder-case.html.

Sutherland, Tara D., James H. Young, Sarah Weisman, Cheryl Y. Hayashi, and David J. Merritt. "Insect Silk: One Name, Many Materials." *Annual Review of Entomology* 55 (Jan. 1, 2010): 171–188.

Szent-Györgyi, Albert. "What is Life?" In *Biology Today*. Del Mar, Calif.: CRM Books, 1972.

Tajfel, Henri. "Experiments in Intergroup Discrimination." *Scientific American* 233, No. 5 (Nov. 1970): 96–103.

Talmage, James E. *The Articles of Faith*. Salt Lake City: Deseret News, 1919.

Taylor, Christopher, and Daniel Dennett. "Who's Afraid of Determinism? Rethinking Causes and Possibilities." In *The Oxford Handbook of Free Will*, ed. Robert Kane, 257–277. Oxford, UK: Oxford University Press, 2005.

Theiler, Anne M., and Louis G. Lippman. "Effects of Mental Practice and Modeling on Guitar and Vocal Performance." *Journal of General Psychology* 122, No. 4 (1999): 329–343.

"This Day in History, July 10: Scopes Monkey Trial Begins." *History*, Nov. 24, 2009. https://www.history.com/this-day-in-history/monkey-trial-begins.

Thomas, Lewis. "On the Uncertainty of Science." *The Key Reporter of Phi Beta Kappa* 46, No. 1 (Autumn 1980).

Thomas, R.D.K., and W.-E. Reif. "The Skeleton Space: A Finite Set of Organic Designs." *Evolution* 47, No. 2 (April 1993): 341–356.

Toledo, Chelsea, and Kirstie Saltsman. "Genetics by the Numbers," National Institute of General Medical Sciences, June 12, 2012. https://www.nigms.nih.gov/education/Inside-Life-Science/Pages/genetics-by-the-numbers.aspx.

Tomasello, Michael. *Why We Cooperate.* Cambridge, Mass.: MIT Press, 2009.

Tomasi, Thomas E. "Echolocation by the Short-Tailed Shrew Blarina brevicauda." *Journal of Mammalogy* 60, No. 4 (Nov. 20, 1979): 751–759.

Tooby, John, and Leda Cosmides. "Friendship, and the Banker's Paradox: Other Pathways to the Evolution of Adaptations for Altruism." *Proceedings of the British Academy* 88 (Jan. 1996): 119–143.

Townes, Charles. "The Convergence of Science and Religion." *Think* 32, No. 2 (March–April 1966).

Trivers, Robert. "The Evolution of Reciprocal Altruism." *Quarterly Review of Biology* 46, No. 1 (March 1971): 35–57.

Tupy, Marian L., and Gale L. Pooley. *Superabundance: The Story of Population Growth, Innovation, and Human Flourishing on an Infinitely Bountiful Planet.* Washington: Cato Institute, 2022.

United Nations Office on Drugs and Crime. *Global Study on Homicide: Executive Summary.* Vienna, 2019. https://www.unodc.org/documents/data-and-analysis/gsh/Booklet1.pdf.

United States Department of Justice. *Correctional Population in the United States, 2010.* December 2011. https://bjs.ojp.gov/content/pub/pdf/cpus10.pdf.

United States Department of Justice. *Crimes in the Unites States, 2011.* Table 42, accessed May 5, 2022. https://ucr.fbi.gov/crime-in-the-u.s/2011/crime-in-the-u.s.-2011/tables/table-42.

United States Food and Drug Administration. Label for Testosterone Gel 1.62%. Accessed May 17, 2022. https://www.accessdata.fda.gov/drugsatfda_docs/label/2017/205781Orig1s000lbl.pdf.

United States v. Grayson, 438 U.S. 41, 1978.

Vaillant, George. *Adaptation to Life.* Cambridge, Mass.: Harvard University Press, 1977.

VanderWeele, Tyler J. "What the *New York Times* Gets Wrong About Marriage, Health, and Well-Being." *Institute for Family Studies,* May 30, 2017, https://ifstudies.org/blog/what-the-new-york-times-gets-wrong-about-marriage-health-and-well-being.

von Ditfurth, Hoimar. *Origins of Life: Evolution as Creation.* Translated by Peter Heinegg. San Francisco: Harper & Row, 1981.

Waite, Linda, and Maggie Gallagher. *The Case for Marriage: Why Married People Are Happier, Healthier, and Better Off Financially.* New York: Broadway Books, 2000.

Wald, George. "Fitness in the Universe: Choices and Necessities." *Origins of Life* 5, No. 1 (Jan.-April 1974): 7–27.

Waldinger, Robert. "What Makes a Good Life? Lessons from the Longest Study on Happiness." Filmed Nov. 2015, TEDx Video: https://www.ted.com/talks/robert_waldinger_what_makes_a_good_life_lessons_from_the_longest_study_on_happiness?language=en.

Wallach, Michael, and Lise Wallach. "How Psychology Sanctions the Cult of the Self." *The Washington Monthly*, February 1985.

Wang, Wendy. "Parents' Time with Kids More Rewarding Than Paid Work—And More Exhausting." *Pew Research Center*, Oct. 8, 2013. https://www.pewresearch.org/social-trends/2013/10/08/parents-time-with-kids-more-rewarding-than-paid-work-and-more-exhausting/.

Warneken, Felix. "How Children Solve the Two Challenges of Cooperation." *Annual Review of Psychology* 69 (Jan. 2018): 205–229.

Warneken, Felix, Brian Hare, Alicia P. Melis, Daniel Hanus, and Michael Tomasello. "Spontaneous Altruism by Chimpanzees and Young Children." *PLoS Biology* 5 (2007): 1414–1420.

Warneken, Felix, and Michael Tomasello. "Altruistic Helping in Human Infants and Young Chimpanzees." *Science* 311, No. 5765 (March 3, 2006): 1301–1303.

Warneken, Felix, and Michael Tomasello. "Extrinsic Rewards Undermine Altruistic Tendencies in 20-month-olds." *Developmental Psychology*, No. 6 (2008): 1785–1788.

Warneken, Felix, and Michael Tomasello. "Helping and Cooperation at 14 Months of Age." *Infancy* 11, No. 3 (2007): 271–294.

Watson, James D., and Francis H. Crick. "Molecular Structure of Nucleic Acids: A Structure for Deoxyribose Nucleic Acid." *Nature* 171, No. 4356 (April 25, 1953): 737–738.

Watson, John B. *Psychological Care of Infant and Child.* New York: W. W Norton & Company, 1928.

Wegner, Daniel. *The Illusion of Conscious Will.* Cambridge, Mass.: MIT Press, 2003.

White, Désirée, and Montserrat Rabago-Smith. "Genotype-Phenotype Associations and Human Eye Color." *Journal of Human Genetics* 56, No. 1 (Jan. 2011): 5–7.

"Why You Do What You Do." *Time*, Aug. 1, 1977. https://content.time.com/time/subscriber/article/0,33009,915181-1,00.html.

Wilcox, W. Bradford. "Don't Be a Bachelor: Why Married Men Work Harder, Smarter and Make More Money." *Washington Post*, April 2, 2015. https

://www.washingtonpost.com/news/inspired-life/wp/2015/04/02/dont-be-a-bachelor-why-married-men-work-harder-and-smarter-and-make-more-money/.

Wilcox, W. Bradford. *Why Marriage Matters: Thirty Conclusions from the Social Sciences* (3rd Edition). New York: Institute for American Values, 2011.

Wilcox, W. Bradford, and Kathleen Kline. *Gender and Parenthood: Biological and Social Scientific Perspectives.* New York: Columbia University Press, 2013.

Wilcox, W. Bradford, Joseph Price, and Robert I. Lerman. "Strong Families, Prosperous States: Do Healthy Families Affect the Wealth of States?" The Home Economics Project, a project of the American Enterprise Institute and the Institute for Family Studies, 2015.

Williams, Ben P., Iain G. Johnston, Sarah Covshoff, and Julian M Hibberd. "Phenotypic Landscape Inference Reveals Multiple Evolutionary Paths to C4 Photosynthesis." *eLife* 2 (Sept. 28, 2013): e00961, https://doi.org/10.7554/eLife.00961.

Williams, Dorie Giles. "Gender, Marriage, and Psychological Well-Being." *Journal of Family Issues* 9, No. 4 (1988): 452–468.

Williams, Kipling D., and Steve A. Nida. "Ostracism: Consequences and Coping." *Current Directions in Psychology* 20, No. 2 (2011): 71–75.

Wilson, Chris M., and Andrew J. Oswald. "How Does Marriage Affect Physical and Psychological Health? A Survey of the Longitudinal Evidence." Discussion Paper No. 1610 (May 2005). https://papers.ssrn.com/sol3/papers.cfm?abstract_id=735205.

Wilson, David Sloan. "Clash of Paradigms: Why Proponents of Multilevel Selection Theory and Inclusive Fitness Theory Sometimes (But Not Always) Misunderstand Each Other." *The Evolution Institute*, July 13, 2012. https://thisviewoflife.com/commentary/david-sloan-wilson-clash-of-paradigms-why-proponents-of-multilevel-selection-theory-and-inclusive-fitness-theory-sometimes-but-not-always-misunderstand-each-other/.

Wilson, David Sloan. *Does Altruism Exist?* New Haven, Conn. and London: Yale University Press, 2015.

Wilson, David Sloan, and Edward O. Wilson. "Rethinking the Theoretical Foundation of Sociobiology." *The Quarterly Review of Biology* 82, No. 4 (Dec. 2007): 327–348.

Wilson, Edward O. *On Human Nature.* Cambridge, Mass.: Harvard University Press, 2004.

Wilson, Edward O. *Sociobiology: The New Synthesis.* Cambridge, Mass.: Harvard University Press, 1975.

Wisniewski, David, Robert Deutschländer, and John-Dylan Haynes. "Free Will Beliefs Are Better Predicted by Dualism Than Determinism Beliefs Across Different Cultures." *PLoS One* 14, No. 9 (Sept. 2019): e0221617.

Woolfolk, Robert L., Mark W. Parrish, and Shane M. Murphy. "The Effects of Positive and Negative Imagery on Motor Skill Performance." *Cognitive Therapy and Research* 9, No. 3 (1985): 335–341.

World Health Organization. "Depression Fact Sheet." Sept. 13, 2021. https://www.who.int/news-room/fact-sheets/detail/depression.

World Health Organization. *Violence Against Women Prevalence Estimates, 2018.* Geneva, 2021. https://www.who.int/publications/i/item/9789240022256.

Wrangham, Richard. *The Goodness Paradox: The Strange Relationship Between Virtue and Violence in Human Evolution.* New York: Pantheon Books, 2019.

Wrangham, Richard. "Two Types of Aggression in Human Evolution." *Proceedings of the National Academy of Science* 115, No. 2 (Jan. 9, 2018): 245–253.

Wu, Katherine J. "Spider Silk is Stronger Than Steel. It Also Assembles Itself." *New York Times*, Nov 4, 2020. https://www.nytimes.com/2020/11/04/science/spider-silk-web-self-assembly.html.

Younger, Jarred, Arthur Aron, Sara Parke, Neil Chatterjee, and Sean Mackay. "Viewing Pictures of a Romantic Partner Reduces Experimental Pain: Involvement of Neural Reward Systems." *PLoS ONE* 5, No. 10 (Oct. 2010): https://doi.org/10.1371/journal.pone.0013309.

Zill, Nicholas, Donna Ruane Morrison, and Mary Jo Coiro. "Long-Term Effects of Parental Divorce on Parent-Child Relationships, Adjustment, and Achievement in Young Adulthood." *Journal of Family Psychology* 7, No. 1 (1993): 91–103.

Notes

Chapter 1: Science, Religion, and the Meaninglessness of Existence

1 Noah Adams, "Timeline: Remembering the Scopes Monkey Trial," July 5, 2005, National Public Radio, https://www.npr.org/2005/07/05/4723956/timeline-remembering-the-scopes-monkey-trial.

2 Ibid.

3 "This Day in History, July 10: Scopes Monkey Trial Begins," *History*, accessed May 3, 2022, https://www.history.com/this-day-in-history/monkey-trial-begins.

4 Ibid.

5 Edward J. Larson, *Summer for the Gods: The Scopes Trial and America's Continuing Debate Over Science and Religion* (New York: Basic Books, 1997), 148.

6 Ibid., 108.

7 Adams, "Timeline."

8 The core of the defense strategy was to show that Scopes's teaching of evolution was not in contradiction to the Bible, and thus did not violate the Butler Act. To do this, they planned to call several scientists who accepted evolution but were also devout Christians.

9 This conclusion of the Scopes Monkey Trial would be correct from the perspective of most lay people. Historian Edward Larson would say that it was almost a draw—both Bryan and Darrow succeeded in turning evolution into a cultural hot-button issue (which is what they intended), though neither really won many converts from the opposite side of the debate. Larson, personal interview, Dec. 3, 2021.

10 Another potential reason was that cracks had been observed in the ceiling below the courtroom, possibly due to a combination of the heat and the record crowds that packed the courtroom throughout the trial.

11 In reality, Darrow had urged the jury to return a guilty verdict. This move was intended to prohibit Bryan from having the last word and delivering an anti-evolution speech for which Bryan had been studiously preparing. See Larson, *Summer for the Gods*, 191.
12 Larson, *Summer for the Gods*, 200.
13 Thomas Henry Huxley to Charles Darwin, November 23, 1859, in The Darwin Correspondence Project, "Letter no. 2544," accessed on September 21, 2021, https://www.darwinproject.ac.uk/letter/DCP-LETT-2544.xml.
14 See Francis Collins, *The Language of God* (New York: Free Press, 2006); Kenneth Miller, *The Human Instinct: How We Evolved to Have Reason, Consciousness, and Free Will* (New York: Simon & Schuster, 2018); Henry Eyring, *The Faith of a Scientist* (Salt Lake City: Bookcraft, 1967); and Charles Townes, "The Convergence of Science and Religion," *Think* 32, No. 2 (March–April 1966).
15 Edward O. Wilson, *On Human Nature* (Cambridge, Mass.: Harvard University Press, 2004 [original edition 1978]), 2.
16 Perhaps one of my favorite quotes in this regard comes from the late Hugh Nibley: "The scientists of past decades have been proud of . . . the 'majestic meaninglessness' of it all. . . . The[se] wise men gloried in the strength of mind and character that enabled them to look an utterly indifferent universe in the face without flinching (after all, they had tenure), insisting that the rest of us rid ourselves of our infantile longings for more." Hugh Nibley, *Old Testament and Related Studies* (Salt Lake City: Deseret Book Company, 1986), 51–52.
17 Tyndal, as quoted in Nibley, *Old Testament and Related Studies*, 52.
18 World Health Organization: Depression Fact Sheet, September 13, 2021, https://www.who.int/news-room/fact-sheets/detail/depression.
19 Anne Case and Angus Deaton. *Deaths of Despair and the Future of Capitalism* (Princeton, N.J., and Oxford, UK: Princeton University Press, 2020).
20 Centers for Disease Control and Prevention. WISQARS Fatal Injury Reports, National, Regional and State, 1981–2018, accessed on September 23, 2021, https://webappa.cdc.gov/sasweb/ncipc/mortrate.html.
21 Richard Dawkins was one of the first to introduce the idea that genes are "replicators," entities that are good at self-copying. He is one of the most ardent proponents of the gene's-eye view of evolution.
22 Within each cell of the human body, an enormous amount of information is condensed into neatly packed chromosomes. If the genes, or DNA, that make up these chromosomes were stretched out linearly, the amount from each cell would measure two to three meters. If the DNA from each cell were lined up end to end, it would stretch beyond Pluto.

23 Lewis Thomas, "On the Uncertainty of Science," *The Key Reporter of Phi Beta Kappa* 46, No. 1 (Autumn 1980), 10.
24 This line is originally from Tennyson's poem, "In Memoriam A.H.H.," originally published in 1850.
25 William Jennings Bryan, *The Prince of Peace* (New York and London: Funk and Wagnalls Company, 1914), 18–19.
26 Richard Dawkins. *The Selfish Gene* (Oxford, UK: Oxford University Press, 1989 [first edition 1976]), 3–4.
27 As one example, see Elizabeth W. Dunn, Lara B. Aknin, and Michael I. Norton, "Spending Money on Others Promotes Happiness," *Science* 319, No. 5870 (March 21, 2008), 1687–1688. Other examples are discussed in Chapter 4.
28 Lara B. Aknin, J. Kiley Hamlin, and Elizabeth W. Dunn, "Giving Leads to Happiness in Young Children," *PLoS ONE* 7, No. 6 (June 2012), https://doi.org/10.1371/journal.pone.0039211.
29 Simon Conway Morris, "The Predictability of Evolution: Glimpses Into a Post-Darwinian World," *Naturewissenschaften* 96, No. 11 (November 2009), 1326.
30 See Martin Seligman, *Flourish: A Visionary New Understanding of Happiness and Well-being* (New York: Free Press, 2011). Also, Harry T. Reis and Shelly L. Gable, "Toward a Positive Psychology of Relationships," in *Flourishing: Positive Psychology and the Life Well-Lived*, eds. C. L. M. Keyes and J. Haidt (American Psychological Association: 2003), 129–159.
31 See David Blankenhorn, *Fatherless America* (New York: Harper Perennial, 1995); David Popenoe, *Life Without Father: Compelling New Evidence That Fatherhood and Marriage Are Indispensable for the Good of Children and Society* (Cambridge, Mass.: Harvard University Press: 1999).
32 Blankenhorn, *Fatherless America*, 25.
33 Theodosius Dobzhansky, "Nothing in Biology Makes Sense Except in the Light of Evolution," *American Biology Teacher* 35, No. 3 (March 1973), 127.
34 This doesn't mean that one cannot believe in miracles, even the impressive miracles described in scripture. We shouldn't necessarily conclude that some of these miracles had to "break" natural laws for them to occur. For instance, the miracle of the parting of the Red Sea is one of the most impressive in the Old Testament. But if you were to debate the merits of this story with a secularist, he would argue that such a story defies the law of gravity and therefore must not be true. The law of gravity, he would assert, does not permit water to behave in this way. But suppose I were to record a thirty-second video of an airplane taking off and I then stored this video on my phone. Suppose I then managed to travel back in time to the 18th century,

with phone in hand, and I began showing the video to individuals of that time period. The fact that I was in possession of such a device as a smartphone would be miraculous enough for anyone who viewed the video from this time period. But suppose the viewer was able to overcome the initial shock of the marvels of a smartphone and focus on the video of the airplane. He or she would rightly be astonished at how a huge piece of metal shaped like an airplane seemingly defies the law of gravity. But of course, we know that there is a natural and logical explanation to this phenomenon. The law of gravity was not broken, but there were other laws and principles at work, namely Bernoulli's principles of fluid dynamics. I disagree with the definition that a miracle must be an event that is not explicable by natural or scientific laws (although some miracles undoubtedly fit this definition according to our current understanding of the laws of nature). One scholar expressed this well: "Miracles are commonly regarded as supernatural occurrences, taking place in opposition to the laws of nature. Such a conception is plainly erroneous, for the laws of nature are inviolable. However, as human understanding of these laws is at best but imperfect, events strictly in accordance with natural law may appear contrary thereto. The entire constitution of nature is founded on system and order; the laws of nature, however, are graded as are the laws of man. The application of a higher law in any particular case does not destroy the efficacy or validity of an inferior one; the lower law is as fully applicable as before to the case for which it is framed." See James E. Talmage, *The Articles of Faith* (Salt Lake City: The Deseret News, 1919), 222.

35 As quoted in Robert Sanders, "UC Berkeley Nobelist Charles Townes to Receive $1.5 Million Templeton Prize for Linking Religion and Science," *UC Berkeley News*, March 9, 2005, https://www.berkeley.edu/news/media/releases/2005/03/09_templetonaward.shtml.

Chapter 2: Evolution: Random Chance or Guided Process?

1 There are various versions of this quote that have been repeated in literature and it's unclear to me where it originated. I use this one from Diane Ackerman, *A Natural History of Love* (New York: Vintage Books, 1995), 260.

2 The late biologist Stephen Jay Gould is generally credited as the one who proposed the thought experiment of what would happen if we rewound time and watched evolution unfold again. According to Gould, the process of evolution was so delicate and subject to chance that it would be highly unlikely that we would arrive at a similar biological world were it to happen again. In fact, Gould suggests that the nonexistence of humans would almost

be a certainty. See his book, *Wonderful Life: The Burgess Shale and the Nature of History* (New York: W.W. Norton & Company, 1989).

3 In my view, I think it's not a given that all life descended from a single cell. The basic building blocks of cells (amino acids, nucleic acids) have been shown to spontaneously form rather easily under the right conditions. This was clearly demonstrated by the Stanley Miller experiment and the discovery of the Murchison meteorite. Given the ubiquity of convergence, even among subcellular machinery, it is possible that there were many original cells. As one scientist notes: "It's obvious that on account of the peculiarities of its atomic structure matter is inclined, 'prefers,' to unite, at every possible opportunity, into compounds we have come to know as the molecules of life. Their spontaneous coming into existence is no longer mysterious; given the influence of the natural laws governing the behavior of matter, the emergence of life seems to have been altogether necessary and inevitable. It should be immediately noted that this in no way makes the whole process less wonderful. The mystery still remains, for how could we ever manage to explain why matter is so constituted that it necessarily produces the chemical conditions required for life?" Hoimar von Ditfurth, *Origins of Life: Evolution as Creation*, transl. Peter Heinegg (San Francisco: Harper & Row, 1981), 46. For reading on the Stanley Miller experiment, see Stanley L. Miller, "A Production of Amino Acids Under Possible Primitive Earth Conditions," *Science* 117, No. 3046 (May 15, 1953), 528–529. For reading on the Murchison meteorite, see Philippe Schmitt-Kopplin, Zelimir Gabelica, Régis D. Gougeon, Agnes Fekete, Basem Kanawati, Mourad Harir, Istvan Gebefuegi, Gerhard Eckel, and Norbert Hertkorn, "High Molecular Diversity of Extraterrestrial Organic Matter in Murchison Meteorite Revealed 40 Years After Its Fall," *Proceedings of the National Academy of Sciences* 107, No. 7 (February 2010): 2763–2768.

4 I should point out here that bird and bat wings are homologous, meaning that their structures derive from a common embryonic origin. This common ancestor, however, did not use its appendages to fly. Later in evolutionary time, birds and bats each independently morphed their appendages into wings that allowed them to fly. The wings of birds and butterflies, however, are analogous, meaning that although they have similar functions, they are of fundamentally different origin. The common node in Figure 2.2 from which the lineages of birds, bats, and butterflies emerge is simplified for the purpose of clarity and is not technically accurate. Birds and bats are more closely related to each other and thus have a more recent shared node.

5 As the biologist George McGhee describes, "The evolution of . . . [the dolphin body shape] is not trivial. It can be correctly described as nothing less than astonishing that a group of land-dwelling [animals], complete with four legs and a tail, could devolve their appendages and their tails back into fins like those of a fish. Highly unlikely, if not impossible? Yet it happened *twice*, convergently in the reptiles and the mammals, two groups of animals that are not closely related." Here, McGhee refers to the fact that not only have porpoises evolved from land-dwelling animals to mimic the structure of sharks (whose ancestors never left the water), but ichthyosaurs (a dolphin-shaped reptile that lived at the time of the dinosaurs) also seemed to have traveled this evolutionary path. George McGhee, "Convergent Evolution: A Periodic Table of Life?" in *The Deep Structure of Biology*, ed. Simon Conway Morris (West Conshohocken, Penn.: Templeton Press, 2008), 19.

6 Simon Conway Morris, *Life's Solution: Inevitable Humans in a Lonely Universe* (Cambridge, UK: Cambridge University Press, 2003), 181.

7 Travis Park, Erich M. G. Fitzgerald, and Alistair R. Evans, "Ultrasonic Hearing and Echolocation in the Earliest Toothed Whales," *Biology Letters* 12 (April 1, 2016), https://doi.org/10.1098/rsbl.2016.0060.

8 Conway Morris, *Life's Solution*, 181.

9 Thomas E. Tomasi, "Echolocation by the Short-Tailed Shrew *Blarina brevicauda*," *Journal of Mammalogy* 60, No. 4 (Nov. 20, 1979), 751–759.

10 Katherine J. Wu, "Spider Silk is Stronger Than Steel. It Also Assembles Itself," *New York Times*, Nov. 4, 2020, https://www.nytimes.com/2020/11/04/science/spider-silk-web-self-assembly.html. Accessed on Sept. 22, 2021.

11 Tara D. Sutherland, James H. Young, Sarah Weisman, Cheryl Y. Hayashi, and David J. Merritt, "Insect Silk: One Name, Many Materials," *Annual Review of Entomology* 55 (Jan. 1. 2010), 171–188.

12 Steven Pinker and Paul Bloom, "Natural Language and Natural Selection," *Behavioral and Brain Sciences* 13 (1990), 709–710.

13 These are known as the polychaetes.

14 See Simon Conway Morris, *Why Are We Here?* Accessed on September 23, 2021, transcript available at https://www.whyarewehere.tv/people/simon-conway-morris/.

15 Russell D. Fernald, "Evolving Eyes," *International Journal of Developmental Biology* 48 (2004), 701–705. Indeed, this estimate of eyes evolving independently between forty and sixty-five times, which is often quoted in the literature, is usually traced back to an older article from 1977 by Luitfried von Salvini-Plawen and Ernst Mayr and is probably outdated (meaning that there are likely more examples of the convergence of eye structures).

16 Richard Dawkins, *Ancestor's Tale: A Pilgrimage to the Dawn of Evolution*, (Boston and New York: Houghton Mifflin, 2004), 588.

17 The process of photosynthesis happens within the cell and involves an extraordinarily complex series of chemical reactions that transforms carbon dioxide (with the help of sunlight) into oxygen and sugar. The usual process is referred to as C3 photosynthesis. But in response to shifts in the environment over the last several million years, plants have ingeniously developed another way to carry out photosynthesis, called C4 photosynthesis. C3 photosynthesis is so named because the first compound formed in the process is comprised of a backbone of three carbon atoms (phosphoglycerate). C4 photosynthesis is so named because the first compound in its process has four carbon atoms, (oxaloacetate). Ben P. Williams, Iain G. Johnston, Sarah Covshoff, and Julian M Hibberd, "Phenotypic Landscape Inference Reveals Multiple Evolutionary Paths to C4 Photosynthesis," e*Life* (Sept. 28, 2013), https://elifesciences.org/articles/00961. It is interesting to note that some scientists think that the evolution of photosynthesis in general is inevitable given similar conditions as have existed on earth. Writing about the possibility of life on other planets, Nobel laureate George Wald remarked that whenever the time comes for a life-form to extract energy from the light emanating from its nearest star, "it seems to me likely that the same factors that governed . . . photosynthesis on the Earth might prove equally compelling elsewhere." In other words, if we find life on other planets (or alternatively, if we could somehow rerun the evolutionary film on our own planet), nature might have developed the same photosynthetic pathways that we have studied in the various plants that populate earth. See George Wald, "Fitness in the Universe: Choices and Necessities," *Origins of Life* 5, No. 1 (Jan.–April 1974), 13–14.

18 At least a few scientists have recognized that the commonplace nature of convergence threatens our scientific confidence in the integrity of phylogenetic trees that evolutionists have constructed. Indeed, it would be much cleaner and easier to construct a great tree of life if convergence did not exist. The development of a phylogenetic tree comes about as scientists recognize that groups of some organisms have more in common than groups of other organisms. The more characteristics that two species share, the more closely related they likely are. It becomes problematic, of course, if organisms that are distantly related converge upon similar characteristics or functions. As Simon Conway Morris states: "I believe the topic of convergence is important for two main reasons. One is widely acknowledged . . . [and] concerns phylogeny, with the obvious circularity of two questions: do we trust our phylogeny and thereby define convergence (which everyone does), or do

we trust our characters to be convergent (for whatever reason) and define our phylogeny? As phylogeny depends on characters, the two questions are inseparable. . . . Even so, no phylogeny is free of its convergences, and it is often the case that a biologist believes a phylogeny because in his or her view certain convergences would be too incredible to be true. . . . During my time in the libraries I have been particularly struck by the adjectives that accompany descriptions of evolutionary convergence. Words like 'remarkable,' 'striking,' 'extraordinary,' or even 'astonishing' and 'uncanny' are commonplace . . . the frequency of adjectival surprise associated with descriptions of convergence suggests there is almost a feeling of unease in these similarities. Indeed, I strongly suspect that some of these biologists sense the ghost of teleology looking over their shoulders." From Conway Morris, *Life's Solution*, 127–128. Nowadays, with our understanding of genetics and genetic mutation as the driving force that underlies evolutionary change, the similarity of gene sequences can help in reconstructing these phylogenetic trees of life. The problem is that there are an increasing number of convergences of molecular biology and even genetic sequences. As Raff, Marshall, and Turbeville have stated: "Given the difficulties associated with alignment and with the establishing the conditions of consistency and convergence, it is clear that molecular phylogenies should not be accepted uncritically as accurate representations of the degree of relatedness between organisms." From Rudolph Raff, Charles R. Marshall, and James M. Turbeville, "Using DNA Sequences to Unravel the Cambrian Radiation of the Animal Phyla," *Annual Review of Ecology and Systematics* 25 (1994), 355. See also Uwe-G Maier, Stefan Zauner, Christian Woehle, Kathrin Bolte, Franziska Hempel, John F. Allen, and William F. Martin, "Massively Convergent Evolution for Ribosomal Protein Gene Content in Plastic and Mitochondrial Genomes," *Genome Biology and Evolution* 5, No. 12 (2013), 2318–2329. My thesis here is not that evolution is false; on the contrary, a great wealth of data suggests that many of the basic principles that underlie evolutionary theory are irrefutable. But the open question is whether evolution was random (even at the genetic and molecular levels), and increasingly the answer seems to be a resounding "no."

19 Conway Morris, *Why Are We Here?* Accessed Sept. 23, 2021.

20 Dawkins, *Ancestor's Tale*, 592.

21 Connie Barlow, "Let There Be Sight! A Celebration of Convergent Evolution," April 2003, http://www.thegreatstory.org/convergence.html.

22 R.D.K. Thomas and W. E. Reif, "The Skeleton Space: A Finite Set of Organic Designs," *Evolution* 47, No. 2 (April 1993), 342.

23 Conway Morris, "The Predictability of Evolution," 1326.

24 See Pascal-Antoine Christin, Daniel M. Weinreich, and Guillaume Besnard, "Causes and Evolutionary Significance of Genetic Convergence," *Trends in Genetics* 26, No. 9 (Sept. 2010), 400–405. See also Rudolph Raff, Charles R. Marshall, and James M. Turbeville, "Using DNA Sequences to Unravel the Cambrian Radiation of the Animal Phyla," *Annual Review of Ecology and Systematics* 25 (1994), 351–375.

25 There is another large body of literature that may allow for this possibility and indirectly support the notion that mutations were not random (at least not *all* of them). Increasingly, scientists are realizing that evolutionary trees based on genetic mutations (molecular phylogenies) are not consistent with one another. For example, in 1997, scientists James Brown and Ford Doolittle published a review that examined the evolutionary trees of over sixty proteins that are essential to all life. The thinking was that these proteins are so essential to all life that they are very tightly "conserved" in evolution, meaning they change very little from organism to organism. But the little change they do exhibit could yield a way to reconstruct an evolutionary tree. The greater the difference in molecular proteins between two organisms, the more distant these organisms were on the tree of life. Conversely, the more similar proteins between two organisms were, the closer they were together on the tree of life. Brown and Doolittle constructed an individual evolutionary tree for each of these sixty-six essential proteins. The very surprising result was that the trees did not match. There were lots of inconsistencies, with trees sprouting all sorts of different branches in different places. (See James R. Brown and W. Ford Doolittle, "*Archae* and the Prokaryote-to-Eukaryote Transition," *Microbiology and Molecular Biology Reviews* 61, No. 4 (1997), 456–502.) The standard explanation to this data (that attempts to preserve the assumption that the mutations were random) is a phenomenon known as *horizontal gene transfer*. Horizontal gene transfer occurs when genetic information (i.e., DNA) is transferred laterally instead of vertically (i.e., through inheritance). Horizontal gene transfer is a real and observed phenomenon among unicellular organisms such as bacteria. But in my opinion, it strains credulity to believe that horizontal gene transfer is responsible for any appreciable amount of apparent genetic relatedness among large multicellular organisms such as ourselves (which we are required to believe if the assumption of random mutation is to hold). For a great book explaining the principle of horizontal gene transfer and how surprisingly widespread it is now thought to be (not unlike the phenomenon of

evolutionary convergence) see David Quammen, *The Tangled Tree: A Radical New History of Life* (New York: Simon & Schuster, 2018).

26 Richard Dawkins, *Climbing Mount Improbable* (New York and London: W.W. Norton & Company, 1996), 70.

27 Scientists sometimes disagree as to which of these reasons is most important, but in reality, in many instances it's likely a combination of both options.

28 Dawkins, *Climbing Mount Improbable.*

29 This analogy of the Museum of All Possible Shells is also borrowed from Dawkins, *Climbing Mount Improbable.*

30 David M. Raup, "Geometric Analysis of Shell Coiling: General Problems," *Journal of Paleontology* 40, No. 5 (Sept. 1966), 1178–1190.

31 Raup uses the terms *T* (translation), *W* (expansion rate), and *D* (distance from axis). Richard Dawkins uses the terms *flare*, *verm*, and *spire*. I think the terms *size, coiling*, and *elongation*, from Christakis's explanation of this phenomenon are easier to visualize. See Christakis, *Blueprint: Evolutionary Origins of a Good Society* (New York: Little, Brown Spark, 2019), 113–114. See also Dawkins, *Climbing Mount Improbable*, 184–189.

32 It's important to note that the rudimentary set of seven variables devised by Thomas and Reif only describe the general form of skeletal patterns. It says nothing of the intricate details of skeleton shapes. In theory, considering all the theoretical geometric possibilities, the number of detailed skeleton shapes is nearly infinite. See Thomas and Reif, "The Skeleton Space," 341–356.

33 Thomas and Reif, "The Skeleton Space," 342.

34 Dawkins, *Climbing Mount Improbable*, 66.

35 Fred Hoyle, "The Universe: Past and Present Reflections," *Engineering and Science* (Nov. 1981), 12. The ideas about the tornado and the universe full of blind men come from Fred Hoyle, *The Intelligent Universe* (New York: Holt, Rinehart, and Winston, 1983).

36 Albert Szent-Györgyi, "What is Life?" In *Biology Today* (Del Mar, Calif.: CRM Books, 1972), xxix–xxxi.

37 See Dawkins, *Climbing Mount Improbable.*

38 See Dawkins, *The Blind Watchmaker: Why the Evidence of Evolution Reveals a Universe without Design* (New York and London: Norton & Company, 1986).

39 The estimate of five million total species comes from Mark J. Costello, Robert M. May, and Nigel E. Stork, "Can We Name Earth's Species Before They Go Extinct?" *Science*, 339, No. 6118 (Jan. 25, 2013), 413–416. The estimate for one trillion species comes from Jay T. Lennon and Kenneth J. Locey, "More Support for Earth's Massive Microbiome," *Biology Direct* 15, No. 5 (March 4, 2020), https://doi.org/10.1186/s13062-020-00261-8.

40 Bill Bryson, *The Body: A Guide for Occupants* (New York: Doubleday, 2019), 294.
41 Ibid.
42 I'm not arguing that embryonic development follows through the exact same phases as evolution. At one point, this was a serious scientific idea referred to as recapitulation theory that is now largely discarded. My point is that the process of self-organization in both cases is similar at a higher level and reflects a process that is not random.
43 Michael Denton and Craig Marshal, "Protein Folds: Laws of Form Revisited," *Nature* 410, No. 6827 (March 22, 2001), 417.

Chapter 3: Evolution: Our Biological and Psychological Heritage

1 See Wilson, *On Human Nature*, 1.
2 Theodore Dobzhansky, "Biology, Molecular and Organismic," *American Zoologist* 4, No. 4 (Nov. 1964), 443–452, 459.
3 Even today, while many educated people readily agree that evolution is responsible for our anatomy and physiology, there still exists considerable resistance to the notion that evolution had anything to do with shaping our behavior.
4 Charles Darwin, *On the Origin of Species by Means of Natural Selection* (New York: Barnes & Noble Classics, 2004, original version published 1859), 60.
5 As quoted in Richard Leakey and Roger Lewin, *Origins: What New Discoveries Reveal About the Emergence of Our Species and its Possible Future* (New York: E.P. Dutton, 1977), 28.
6 See Herbert Spencer's *Principles of Biology,* 1864. The phrase "survival of the fittest" has been roundly criticized over the years by both scientists and laypeople. Most modern biologists prefer to simply use the term *natural selection*. One aspect that I dislike from the phrase is that survival is only part of the equation. You have to survive long enough to reproduce and, if your species is a k-selected species (like ours), rear your young to reproductive maturity.
7 Jonathan B. Losos, "What Is Evolution?" in *The Princeton Guide to Evolution*, ed. Jonathan B. Losos (Princeton, N.J.: Princeton University Press, 2014), 8.
8 As quoted in Dan I. Andersson, "Evolution of Antibiotic Resistance," in *Princeton Guide to Evolution*, ed. Jonathan B. Losos (Princeton, N.J.: Princeton University Press, 2014), 748.
9 C.C.S. Fuda, J. F. Fisher, and S. Mobashery, "β-Lactam Resistance in *Staphylococcus aureus*: The Adaptive Resistance of a Plastic Genome," *Cellular and Molecular Life Sciences* 62 (Sept. 7, 2005), 2617–2633.
10 This protein is called penicillin-binding protein 2a (PBP 2a).
11 See Bill Bryson, *The Body: A Guide for Occupants* (New York: Doubleday, 2019), 45–46.

12 Pew Research Center. Americans, Politics and Science Issues, July 1, 2015. Chapter 4: Evolution and perceptions of scientific consensus. https://www.pewresearch.org/science/2015/07/01/chapter-4-evolution-and-perceptions-of-scientific-consensus/.

13 See *The Princeton Guide to Evolution*, ed. Jonathan B. Losos, 2017.

14 Charles William Eliot, who served as Harvard's president for forty years, was a vocal supporter of eugenics and forced sterilization of the so-called "feeble minded." See Adam S. Cohen, "Harvard's Eugenics Era," *Harvard Magazine*, March–April 2016, https://www.harvardmagazine.com/2016/03/harvards-eugenics-era.

15 Wilson, *On Human Nature*, 1.

16 James D. Watson and Francis H. Crick, "Molecular Structure of Nucleic Acids: A Structure for Deoxyribose Nucleic Acid," *Nature* 171, No. 4356 (April 25, 1953), 737–738.

17 Chelsea Toledo and Kirstie Saltsman, "Genetics by the Numbers," National Institute of General Medical Sciences, June 12, 2012, https://www.nigms.nih.gov/education/Inside-Life-Science/Pages/genetics-by-the-numbers.aspx.

18 Anna Bębenek and Izabela Ziuzia-Graczyk, "Fidelity of DNA Replication—A Matter of Proofreading," *Current Genetics* 64 (2018), 985–996.

19 Of course, as discussed in Chapter 2, the phenomena of convergent evolution (even among genetic components) and horizontal gene transfer (which purportedly happened frequently even in multicellular organisms) casts considerable doubt on the overall structure and relatedness of life as a tree. See Conway Morris, *Life's Solution* and Quammen, *The Tangled Tree.*

20 See Bryson, *The Body*, 7.

21 James F. Crow, "Unequal by Nature: A Geneticist's Perspective on Human Differences," *Daedalus: Journal of the American Academy of Arts & Sciences* 131, No. 1 (Winter 2002), 81.

22 Bonobos are another primate species that are often said to be the most genetically similar to humans.

23 Kay Prüfer, Kasper Munch, Ines Hellmann, Keiko Akagi, James R. Miller, Brian Walenz, et al., "The Bonobo Genome Compared with the Chimpanzee and Human Genomes," *Nature* 486, No. 7404 (June 28, 2012), 527–531.

24 Joan U. Pontius, James C. Mullikin, Douglas R. Smith, Agencourt Sequencing Team, Kerstin Lindblad-Toh, et al., "Initial Sequence and Comparative Analysis of the Cat Genomes," *Genome Research* 17, No. 11 (Nov. 2007), 1675–1689.

25 "Why Mouse Matters," National Human Genome Research Institute, accessed May 3, 2022, https://www.genome.gov/10001345/importance-of-mouse-genome.

26 International Chicken Genome Sequencing Consortium, "Sequence and comparative analysis of the chicken genome provide unique perspectives on vertebrate evolution," *Nature* 432, No. 7018 (Dec. 9, 2004), 695–777.

27 The truth (as usual) is a bit more nuanced. It turns out that about 60 percent of our genes have recognizable counterpart genes in a banana. Of those genes that have counterparts, they are not identical when we compare the actual sequence of amino acids. The same thing goes for many of the other comparisons. The oft-repeated figure that we share somewhere between 41 and 60 percent of our genes with bananas likely comes from an educational video called *The Animated Genome* produced by the National Human Genome Research Institute (part of the National Institutes of Health) in 2013, led by Dr. Lawrence Brody. See Alia Hoyt, "Do People and Bananas Really Share 50 Percent of the Same DNA?" *HowStuffWorks*, April 1, 2021, https://science.howstuffworks.com/life/genetic/people-bananas-share-dna.htm.

28 See, for instance, Barbara Bradley Hagerty, "Can Your Genes Make You Murder?" *National Public Radio*, July 1, 2010, https://www.npr.org/templates/story/story.php?storyId=128043329. See also Andy Stiny, "'Warrior Gene' Defense Mounted in Santa Fe Murder Case," *Albuquerque Journal*, October 24, 2014, https://www.abqjournal.com/485234/warrior-gene-defense-mounted-in-santa-fe-murder-case.html.

29 Genex Diagnostics, accessed April 26, 2022, https://www.genexdiagnostics.com/.

30 Paul R. Ehrlich, *Human Natures: Genes, Cultures, and the Human Prospect* (New York: Penguin, 2000), 4.

31 Désirée White and Montserrat Rabago-Smith, "Genotype-Phenotype Associations and Human Eye Color," *Journal of Human Genetics* 56, No. 1 (Jan. 2011), 5–7.

32 Elena A. Ponomarenko, Ekaterina V. Poverennaya, Ekaterina V. Ilgisonis, Mikhail A. Pyatnitskiy, Arthur T. Kopylov, Victor G. Zgoda, Andrey V. Lisitsa, and Alexander I. Archakov, "The Size of the Human Proteome: The Width and Depth," *International Journal of Analytical Chemistry* 2016, Article ID 7436849, https://doi.org/10.1155/2016/7436849.

33 Robert Sapolsky, *Behave: The Biology of Humans at Our Best and Worst* (New York: Penguin, 2017), 248.

34 For a great summary of how mind-boggling the complexity of genetics can be, see Chapter 8 from Sapolsky, *Behave*.

35 Elliott Sober and David Sloan Wilson, *Unto Others: The Evolution and Psychology of Unselfish Behavior* (Cambridge, Mass.: Harvard University Press, 1998), 3. In this book, Sober and Wilson distinguish between what they term evolutionary altruism (the behavior) and psychological altruism

(the motivation behind it). I generally won't make a distinction between these two in this book, but there are many reasons someone might behave altruistically.

36 Sober and Wilson, *Unto Others*, 3.

37 Darwin, *The Origin of Species*, 194.

38 Laura Bortolotti and Cecilia Costa, "Chemical Communication in the Honey Bee Society," in *Neurobiology of Chemical Communication*, ed. Carla Mucignat-Caretta (Boca Raton, Fla.: CRC Press/Taylor & Francis, 2014).

39 Darwin, *The Origin of Species*, 79.

40 Edward O. Wilson, *Sociobiology: The New Synthesis* (Cambridge, Mass.: Harvard University Press, 1975), 117.

41 Wilson, *Sociobiology*, 118.

42 Michael R. Cunningham, "Levites and Brother's Keepers: A Sociobiological Perspective on Prosocial Behavior," *Humboldt Journal of Social Relations* 13, No. 1/2 (1986), 35–67.

43 Elaine A. Madsen, Richard J. Tunney, George Fieldman, Henry C. Plotkin, Robin I. M. Dunbar, Jean-Marie Richardson, and David McFarland, "Kinship and Altruism: A Cross-Cultural Experimental Study," *British Journal of Psychology* 98, No. 2 (May 2007), 339–359.

44 S. Bowles and D. Posel, "Genetic Relatedness Predicts South African Migrant Workers' Remittances to Their Families," *Nature* 434, No. 7031 (March 17, 2005), 380–383.

45 Maxwell N. Burton-Chellew and Robin I. M. Dunbar, "Hamilton's Rule Predicts Anticipated Social Support in Humans," *Behavioral Ecology* 26, No. 1 (Jan.–Feb. 2015), 130–137.

46 See Nigel Barber, "Machiavellianism and Altruism: Effect of Relatedness of Target Persons on Machiavellian and Helping Attitudes," *Psychological Reports* 75 (1994), 403–422.

47 William M. Muir and David Sloan Wilson, "When the Strong Outbreed the Weak: An Interview with William Muir," *This View of Life*, July 11, 2016, https://thisviewoflife.com/when-the-strong-outbreed-the-weak-an-interview-with-william-muir/.

48 Jonathan N. Pruitt and Charles J. Goodnight, "Site-Specific Group Selection Drives Locally Adapted Group Compositions," *Nature* 514, No. 7522 (Oct. 16, 2014), 359–362.

49 Charles Darwin, *The Descent of Man, and Selection in Relation to Sex* (London: Penguin, 1871 [2nd edition 1879]), 157–158.

50 David Sloan Wilson and Edward O. Wilson, "Rethinking the Theoretical Foundation of Sociobiology," *The Quarterly Review of Biology* 82, No. 4 (Dec. 2007), 345.

51 For example, Samir Okasha argues that group selection in some form must have happened as multicellular organisms developed (which happened many times due to convergent evolution, see Chapter 2). At some point, groups of cells coalesced and started working together harmoniously and started evolving as a unit. This is how multicellular organisms first developed. See Samir Okasha, *Philosophy of Biology: A Very Short Introduction* (Oxford, UK: Oxford University Press, 2019).

52 See Samuel Bowles, "Group Competition, Reproductive Leveling, and the Evolution of Human Altruism," *Science* 314, No. 5805 (Dec. 8, 2006), 1569–1572. See also Samuel Bowles, "Did Warfare Among Ancestral Hunter-Gatherers Affect the Evolution of Human Social Behaviors?" *Science* 324, No. 5932 (June 5, 2009), 1293–1298.

53 The interested reader is referred to an essay by David Sloan Wilson that helps strike at the core of this debate and shows how, in some ways, critics of either side may be arguing about semantics more than substance. See David Sloan Wilson, "Clash of Paradigms: Why Proponents of Multilevel Selection Theory and Inclusive Fitness Theory Sometimes (But Not Always) Misunderstand Each Other," *The Evolution Institute*, July 13, 2012, https://thisviewoflife.com/commentary/david-sloan-wilson-clash-of-paradigms-why-proponents-of-multilevel-selection-theory-and-inclusive-fitness-theory-sometimes-but-not-always-misunderstand-each-other/. See also Sober and Wilson, *Unto Others*. In Chapter 5, Sober and Wilson report what they claim amounts to a "smoking gun" of group selection: the phenomenon occurring before their eyes. This example involves a long-standing conflict between two African tribes near modern Sudan: the Nuer and the Dinka. According to Sober and Wilson, the Nuer culture is slowly displacing the Dinka culture, in part because the Nuer are more cohesive as a group. Here (as is often understood to be the case for group selection), the process is more of a cultural evolution than a biological or genetic one (the tribes frequently intermarry).

54 Stephen C. Stearns, "Natural Selection, Adaptation, and Fitness: Overview," in *Princeton Guide to Evolution*, ed. Jonathan B. Losos (Princeton, N.J.: Princeton University Press, 2014), 193–199.

55 W. D. Hamilton, "Innate Social Aptitudes of Man: An Approach from Evolutionary Genetics," in *ASA Studies 4: Biosocial Anthropology*, ed. Robin Fox (London: Malaby Press, 1975), 332–333.

Chapter 4: Selfishness and Altruism

1 This quote is traditionally attributed to Roman emperor Marcus Aurelius.

2 Brian Cruver, *Anatomy of Greed: The Unshredded Truth from an Enron Insider* (New York: Carol & Graf Publishers, 2002), 88.
3 Ibid., 61.
4 Ibid., 63.
5 Ibid.
6 This is in line with the self-domestication hypothesis developed (in part) by Richard Wrangham. See *The Goodness Paradox: The Strange Relationship Between Virtue and Violence in Human Evolution* (New York: Pantheon Books, 2019). Wrangham argues that humans underwent a sort of self-domestication process where we ostracized or even put to death extremely selfish people and that this is one of the reasons that accounts for why we are so cooperative.
7 The importance of punishment and the development of guilt, shame, and remorse were likely all critical to keeping would-be cheaters from dominating society. The biologist Robert D. Alexander remarked: "Obviously, various forms of punishment, including ostracism or social shunning, can . . . be applied to individuals repeatedly observed not to reciprocate adequately or follow whatever codes of conduct may exist." See Robert D. Alexander, *The Biology of Moral Systems* (New York: Adline de Gruyter, 1987), 94.
8 Wilson, *Sociobiology*, 122.
9 See, for instance, Robert M. Seyfarth, Dorothy L. Cheney, and Peter Marler, "Vervet Monkey Alarm Calls: Semantic Communication in a Free-Ranging Primate," *Animal Behaviour* 28, No. 4 (1980), 1070–1094.
10 For his original analysis, see Nicholas Christakis, *Blueprint: Evolutionary Origins of a Good Society* (New York: Little, Brown Spark, 2019), 38–44. Here I retell the stories in different words drawing upon original detailed sources, while continuing to be influenced by the analysis of Christakis.
11 Christakis, *Blueprint*, 42.
12 Joan Druett, *Island of the Lost: An Extraordinary Story of Survival at the Edge of the World* (Chapel Hill, N.C.: Algonquin Books, 2007), 37.
13 Ibid., 45.
14 Ibid., 48.
15 Ibid., 49.
16 Ibid., 73.
17 Ibid., 75.
18 Ibid., 209.
19 Ibid.
20 Ibid., 236.
21 Madelene Allen, *Wake of the Ivercauld: Shipwrecked in the Sub-Antarctic: A Great-Granddaughter's Pilgrimage* (Auckland, NZ: Exisle Publishing, 1997), 242.

22 Ibid., 248.
23 Ibid., 249.
24 Ibid.
25 Druett, *Island of the Lost*, 199.
26 Ibid., 251.
27 Ibid., 249.
28 Here, again, I draw upon the analysis and conclusions of Christakis in comparing these events.
29 See again Christakis's analysis of these events, *Blueprint*, 44.
30 Christakis, *Blueprint*, 42.
31 For example, during World War II, seventeen-year-old private Jack Lucas saved three comrades when he jumped on top of two grenades that had been thrown into his foxhole. Remarkably, he survived the ordeal. Desmond Doss risked his life to save approximately seventy-five of his comrades who were wounded in the Battle of Okinawa. Both Lucas and Doss were awarded the Medal of Honor by President Harry Truman.
32 Nicole Chavez and Marlena Baldacci, "This UNC Charlotte Student Knocked the Shooter Off His Feet and Saved Lives," CNN, May 2, 2019, https://www.cnn.com/2019/05/01/us/uncc-victims-shooting-riley-howell/index.html, accessed June 29, 2021.
33 Trevor Hughes and Rebecca Plevin, "Her 'Final Good Deed': Woman Hailed as Hero After Taking Bullets to Protect Rabbi During Synagogue Shooting," *USA Today*, April 28, 2019, https://www.usatoday.com/story/news/2019/04/28/lori-gilbert-kaye-woman-protected-rabbi-synagogue-shooting/3608418002/, accessed June 29, 2021.
34 Bill Chappell and Sasha Ingber, "'This is Not Who We Are,' Colorado Officials Say After Deadly School Shooting," NPR, May 8, 2019, https://www.npr.org/2019/05/08/721474989/this-is-not-who-we-are-colorado-officials-say-after-deadly-school-shooting, accessed June 29, 2021.
35 Wilson, *On Human Nature*, 150.
36 Felix Warneken and Michael Tomasello, "Altruistic Helping in Human Infants and Young Chimpanzees," *Science* 311, No. 5765 (March 3, 2006), 1301–1303.
37 See, for instance, Felix Warneken and Michael Tomasello, "Helping and Cooperation at 14 Months of Age," *Infancy* 11, No. 3 (2007), 271–294. See also Kristen A. Dunfield and Valerie A. Kuhlmeier, "Intention-mediated Selective Helping in Infancy," *Psychological Science* 21, No. 4 (April 2010), 523–527. For a review, see Michael Tomasello, *Why We Cooperate* (Cambridge, Mass.: MIT Press, 2009) or Felix Warneken, "How Children Solve the Two Challenges of Cooperation," *Annual Review of Psychology* 69 (Jan. 2018), 205–229.

38 Christopher Boehm argues that a "slippage model" or spillover effect may partially account for why humans are so altruistic toward nonkin. See *Moral Origins: The Evolution of Virtue, Altruism, and Shame* (New York: Basic Books, 2012), 57–58.

39 See David Popenoe, *Life Without Father*, 75.

40 Robert Trivers, "The Evolution of Reciprocal Altruism," *Quarterly Review of Biology* 46, No. 1 (March 1971), 35.

41 Merrill Flood and Melvin Dresher developed the game while working at RAND.

42 Richard Dawkins has an excellent and expanded description of this in *The Selfish Gene*.

43 Robert Axelrod, "Effective Choice in the Prisoner's Dilemma," *Journal of Conflict Resolution* 24, No. 1 (March 1, 1980), 3–25.

44 As David Sloan Wilson has pointed out, reciprocal altruism is actually a special case of group selection. Reciprocal altruism essentially explains how groups of two cooperators outcompete groups of two defectors. The recognition of this goes far to advance the case that group selection has played a role some of the time in evolution.

45 Quoted by Boehm, *Moral Origins*, 166.

46 Incidentally, this phenomenon is not observed in our closest primate relatives: chimpanzees. See Jan M. Engelmann, Esther Herrmann, and Michael Tomasello, "Five-Year-Olds, but Not Chimpanzees, Attempt to Manage Their Reputations," *PloS ONE* 7, No. 10 (Oct. 2010), e48433, https://doi.org/10.1371/journal.pone.0048433. See also Jan M. Engelmann, Esther Herrmann, and Michael Tomasello, "The Effects of Being Watched on Resource Acquisition in Chimpanzees and Human Children," *Animal Cognition* 19 (2016), 147–151. A key part of this likely has to do with language ability. It is very likely that chimpanzees recognize that variation in altruism exists in their species; some chimpanzees are more giving than others. But they cannot readily communicate this with each other, at least not to the same level that humans can. Hence, if reputation cannot be communicated to others, it has little evolutionary or social value.

47 See again Engelmann, Herrmann, and Tomasello, "The Effects of Being Watched," 2016. For a review, see Tomasello, *Why We Cooperate*.

48 Melissa Bateson, Daniel Nettle, and Gilbert Roberts, "Cues of Being Watched Enhance Cooperation in a Real-World Setting," *Biology Letters* 2, No. 3 (June 27, 2006), 412–414.

49 Damien Francey and Ralph Bergmuller, "Images of Eyes Enhance Investments in a Real-Life Public Good," *PloS One* 7, No. 5 (May 2012), https://doi.org/10.1371/journal.pone.0037397.

50 See Christakis, *Blueprint*, 248.

51 See, for example, Bowles, "Did Warfare Among Ancestral Hunter Gatherers Affect Evolution of Human Social Behaviors?" 2009. Bowles argues that the extreme volatility in the climate during the late Pleistocene era (roughly 120,000 to 20,000 years ago) led to conditions that promoted altruism and cooperation (as well as intergroup conflict).

52 John Tooby and Leda Cosmides, "Friendship, and the Banker's Paradox: Other Pathways to the Evolution of Adaptations for Altruism," *Proceedings of the British Academy* 88 (Jan. 1996), 131.

53 See also Christakis's treatment of this subject in *Blueprint*, Chapter 8.

54 Elizabeth W. Dunn, Lara B. Aknin, and Michael I. Norton, "Spending Money on Others Promotes Happiness," *Science* 319, No. 5870 (March 21, 2008), 1687–1688.

55 Lara B. Aknin, J. Kiley Hamlin, and Elizabeth W. Dunn, "Giving Leads to Happiness in Young Children," *PloS ONE* 7, No. 6 (June 2012), https://doi.org/10.1371/journal.pone.0039211.

56 Lara B. Aknin, Tanya Broesch, J. Kiley Hamlin, and Julia W. Van de Vondervoort, "Prosocial Behavior Leads to Happiness in a Small-Scale Rural Society," *Journal of Experimental Psychology: General* 144, No. 4 (Aug. 2015), 788–795.

57 The Acts of the Apostles, 20:35 (KJV).

58 Lara B. Aknin, Alice L. Fleerackers, and J. Kiley Hamlin, "Can Third-Party Observers Detect Emotional Rewards of Generous Spending?" *Journal of Positive Psychology* 9, No. 3 (2014).

59 See Martin Seligman, *Flourish: A Visionary New Understanding of Happiness and Well-being* (New York: Free Press, 2011).

60 Louise C. Hawkley and John T. Cacioppo, "Loneliness Matters: A Theoretical and Empirical Review of Consequences and Mechanisms," *Annals of Behavioral Medicine* 40, No. 2 (July 22, 2010), 218–227. See also Raffaella Calati, Chiara Ferrari, Marie Brittner, Osmano Oasi, Emilie Olié, André F. Carvahlo, and Philippe Courtet, "Suicidal Thoughts and Behaviors and Social Isolation: A Narrative Review of the Literature," *Journal of Affective Disorders* 245 (2019), 653–667.

61 See Julianna Holt-Lunstad, Timothy B. Smith, Mark Baker, Tyler Harris, and David Stephenson, "Loneliness and Social Isolation as Risk Factors for Mortality: A Meta-Analytic Review," *Perspectives on Psychological Science* 10, No. 2 (March 11, 2015), 227–237. See also Cigna US Loneliness Index: Survey of 20,000 Americans Examining Behaviors Driving Loneliness in the United States, May 2018, available at https://www.multivu.com/players

/English/8294451-cigna-us-loneliness-survey/docs/IndexReport_1524069 371598-173525450.pdf.

62 So which is it? Did our altruistic tendencies arise from direct reciprocity? Or was it the reputation hypothesis (indirect reciprocity)? Or was it kin selection, or even a bit of group selection that set the scene for human altruism? The truth is likely a mix of all of the above. These different origins of altruism may correspond roughly to different motivations behind the behavior of human giving. As Christopher Boehm remarks: "You may give to another because you expect an immediate or eventual return; you may give because you fear gossip and public opinion; you may give because seriously unacceptable nongiving can bring active punishment from your peers; you may give as a social conformist just because that's what people do in your culture, and this makes for an easy conscience. And, of course, you may also give in a heartfelt way because you identify with another's need or distress and it just feels right to be helpful." See Boehm, *Moral Origins*, 55.

63 Wilson, *Sociobiology*, 129.

Chapter 5: Aggression and Cooperation

1 Richard Leakey and Roger Lewin, *Origins: What New Discoveries Reveal About the Emergence of Our Species and Its Possible Future* (New York: E.P. Dutton, 1977), 245.

2 Sigmund Freud, *Civilization and Its Discontents*, transl. David McLintock (New York: Penguin, 2002), 56.

3 See Robert Sapolsky, *Behave: The Biology of Humans at Our Best and Worst* (New York: Penguin, 2017).

4 Anthropologist David Carrier argues that our capacity for violence and aggression are biologically etched into our bodies. While most experts would suggest that the anatomy of our hands is adapted to optimize dexterity (for tool building and other complex activities), Carrier suggests that hand anatomy is also adapted for violence. Our hands, he claims, evolved to be efficient weapons used for punching. Carrier and his lab have gone so far as to test the biomechanics of the fists of human cadavers in punching experiments, showing that a buttressed fist could deliver a blow with more force compared to an open-handed slap. See, for instance, Joshua Horns, Rebekah Jung, and David Carrier, "*In vitro* Strain in Human Metacarpal Bones During Striking: Testing the Pugilism Hypothesis of Hominin Hand Evolution," *Journal of Experimental Biology* 218, No. 20 (2015), 3215–3221. Not all anthropologists agree with Carrier's conclusions. It is a far cry, they argue, from showing that the fist is a stronger punching weapon than an open-hand slap to concluding

that the hand evolved by nature selecting violent hominids. It is possible that humans learned to make fists once they were formed, taking advantage of the way hands were shaped for dexterity and using them for fighting. See David C. Nickle and Leda M. Goncharoff, "Human Fist Evolution: A Critique," *Journal of Experimental Biology* 216, No. 12 (June 2013), 2359–2360.

5 See Lee T. Gettler, Thomas W. McDade, Alan B. Feranil, and Christopher W. Kuzawa, "Longitudinal Evidence that Fatherhood Decreases Testosterone in Human Males," *Proceedings of the National Academy of Sciences* 108, No. 39 (Sept. 27, 2011), 16194–16199. Again, the relationship between biology and culture is complicated, and attitude and expectation play a role. In situations where a man has an expectation of fidelity to his partner, testosterone levels are lower than when a man keeps his options open for new sexual prospects. See Matthew McIntyre, Steven W. Gangestad, Peter B. Gray, Judith Flynn Chapman, Terence C. Burnham, Mary T. O'Rourke, and Randy Thornhill, "Romantic Involvement Often Reduces Men's Testosterone Levels—But Not Always: The Moderating Role of Extrapair Sexual Interest," *Journal of Personality and Social Psychology* 91, No. 4 (Oct. 2006), 642–651.

6 David J. Handelsman, Angelica L. Hirschberg, and Stephane Bermon, "Circulating Testosterone as the Hormonal Basis of Sex Differences in Athletic Performance," *Endocrine Reviews* 39, No. 5 (Oct. 2018), 803–829.

7 For a brief review, see K. K. Soma, "Testosterone and Aggression: Berthold, Birds and Beyond," *Journal of Neuroendocrinology* 18, No. 7 (July 2006): 543–544.

8 See Wilson, *Sociobiology*, 251.

9 My thoughts here are influenced by Richard Wrangham's book *Goodness Paradox: The Strange Relationship Between Virtue and Violence in Human Evolution* (New York: Pantheon Books, 2019). Wrangham teaches eloquently about the MAOA gene and its influence on aggressive behavior, tells the story of Adrian Raine, and argues convincingly that biology influences but does not determine aggression. Here I remake this argument with my own words using original sources.

10 Rose McDermott, Dustin Tingley, Jonathan Cowden, Giovanni Frazzetto, and Dominic D. P. Johnson, "Monoamine Oxidase A Gene (MAOA) Predicts Behavioral Aggression Following Provocation," *Proceedings of the National Academy of Sciences* 106, No. 7 (Feb. 17, 2009), 2118–2123.

11 Courtney A. Ficks and Irwin D. Waldman, "Candidate Genes for Aggression and Antisocial Behavior: A Meta-Analysis of Association Studies of the *5HTTLPR* and *MAOA-uVNTR*," *Behavior Genetics* 44, (2014), 427–444.

12 Adrian Raine, J. Reid Meloy, Susan Bihrle, Jackie Stoddard, Lori LaCasse, and Monte S. Buchsbaum, "Reduced Prefrontal and Increased Subcortical

Brain Functioning Assessed Using Positron Emission Tomography in Predatory and Affective Murderers," *Behavioral Sciences & The Law* 16, No. 3 (Summer 1998), 319–332. This study was primarily aimed at distinguishing between reactive and premeditated aggression, discussed later in this chapter. However, men who committed both types of aggression (murder) as examined in this study were sufficiently different in several ways from men who had not done so, most notably overactive subcortical areas in the right hemisphere of the brain. See also Adrian Raine, Monte S. Buchsbaum, and Lori LaCasse, "Brain Abnormalities in Murderers Indicated by Positron Emission Tomography," *Biological Psychiatry* 42, No. 6 (Sept. 15, 1997), 495–508.

13 Lee Ellis, Kevin Beaver, and John Wright, *Handbook of Crime Correlates* (San Diego: Academic Press, 2009), 208–210.

14 Lauren E. Glaze, "Correctional Populations in the United States, 2010," Bureau of Justice Statistics, December 2011, available at https://bjs.ojp.gov/content/pub/pdf/cpus10.pdf.

15 United Nations Office on Drugs and Crime, *Global Study on Homicide: Executive Summary* (Vienna, 2019), 11–17, available at https://www.unodc.org/documents/data-and-analysis/gsh/Booklet1.pdf.

16 Deborah Denno, "Courts' Increasing Consideration of Behavioral Genetics Evidence in Criminal Cases: Results of a Longitudinal Study," *Michigan State Law Review* 967 (2011), 980–981.

17 H. G. Brunner, M. Nelen, X. O. Breakefield, H. H. Ropers, and B. A. Van Oost, "Abnormal Behavior Associated with a Point Mutation in the Structural Gene for Monoamine Oxidase A," *Science* 262, No. 5133 (Oct. 22, 1993), 578–580.

18 Shalender Bhasin, Juan P. Brito, Glenn R. Cunningham, Frances J. Hayes, Howard N. Hodis, Alvin M. Matsumoto, Peter J. Snyder, Ronald S. Swedloff, Frederick C. Wu, and Maria A. Yialamas, "Testosterone Therapy in Male with Hypogonadism: An Endocrine Society Clinical Practice Guidelines," *Journal of Clinical Endocrinology and Metabolism* 103, No. 5 (May 2018), 1715–1744. This practice guideline does not even list aggression as a side effect of testosterone therapy. The label from the Food and Drug Administration (FDA) reports that 2.6 percent of patients taking testosterone gel report some type of "emotional lability," which includes mood swings, impatience, anger, and aggression. See Testosterone Gel 1.62 percent, U.S. Food and Drug Administration website, https://www.accessdata.fda.gov/drugsatfda_docs/label/2017/205781Orig1s000lbl.pdf, accessed May 17, 2022.

19 In one meta-analysis, the correlation of the relationship between testosterone and aggression ranges from r=0.02 to r=0.17, depending upon age, gender,

time of testosterone measurement (levels vary throughout the day), and other factors. This means that, at the very most, about 3 percent of the variation in aggressive behavior can be explained by the variation in testosterone levels. See John Archer, Nicola Graham-Kevan, and Michelle Davies, "Testosterone and Aggression: A Reanalysis of Book, Starzyk, and Quinsey's (2001) Study," *Aggression and Violent Behavior* 10 (2005), 241–261.

20 We also see the fallacy of biology as a deterministic cause of aggression when we return to Adrian Raine's brain scan studies of murderers. When Raine scanned his own brain, it showed a closer resemblance to the brains of the accused murderers than to the healthy control subjects. Raine later commented on this disturbing finding: "When you have a brain scan that looks like a serial killer's it does give you pause." Quoted by Tim Adams in "How to Spot a Murderer's Brain," *Guardian*, May 11, 2013, https://www.theguardian.com/science/2013/may/12/how-to-spot-a-murderers-brain.

21 The neuroendocrinologist Robert Sapolsky reports, "The winning effect on testosterone is lessened in circumstances where people feel . . . like they underperformed. In contrast, it is enhanced among people who went into the competition having the strongest psychological motives for domination. Finally, testosterone levels can rise robustly in 'losers' who nonetheless performed far better than they anticipated. Thus one might see testosterone levels rise after a marathon in a guy who came in at the back of the pack but is triumphant because he was sure he was going to drop dead halfway through, and may decline in the guy who comes in third but was expected to win." See Sapolsky, *Behave*, 105.

22 Wrangham, *Goodness Paradox*, 46.

23 For his definition of combative sport, Sipes defined it as: "one played by two opponents (individuals or teams) and fulfilling one or both of the following conditions: (1) There is actual or potential body contact between opponents, either direct or though real or symbolic weapons. One of the objectives of the sport appears to be inflicting real or symbolic bodily harm on the opponent or gaining playing field territory from the opponent (which would include placing a disputed object in, or acquiring one from, a guarded location). . . . (2) There is no body contact, harm, or territorial gain but there is patently warlike activity. Such sports, to be classified as combative, must include use of actual or simulated combat weapons against an actual or simulated human being." See Richard G. Sipes, "War, Sports and Aggression: An Empirical Test of Two Rival Theories," *American Anthropologist* 75 (1973), 69–70.

24 Sipes, "War, Sports, and Aggression." Notably, there were several flaws of Sipes's cross-cultural examination of war and sports. Among these flaws were

that he chose cultures on the extreme end, deliberately seeking peaceful and warlike societies. A second flaw was that he only examined twenty cultures, a relatively small number. However, the overall finding that aggression is a learned behavior (as opposed to the "drive-discharge" model of aggression), interpreted from the variations of war and combative sports was replicated in a larger set of societies and a more recent and sophisticated analysis. See Garry Chick, John W. Loy, and Andrew W. Miracle, "Combative Sport and Warfare: A Reappraisal of the Spillover and Catharsis Hypotheses," *Cross-Cultural Research* 31, No. 3 (1997), 249–267.

25 See Albert Bandura, Dorothea Ross, and Sheila A. Ross, "Transmission of Aggression Through Imitation of Aggressive Models," *Journal of Abnormal and Social Psychology* 63, No. 3 (1961), 575–582. Other of Bandura's experiments confirm a large amount of learning in aggressive behavior: Albert Bandura, "Influence of Models' Reinforcement Contingencies on the Acquisition of Imitative Responses," *Journal of Personality and Social Psychology* 1, No. 6 (1965), 589–595; Albert Bandura, Dorothea Ross, and Sheila A. Ross, "Imitation of Film-Mediated Aggressive Models," *Journal of Abnormal and Social Psychology* 66, No. 1 (1963), 3–11.

26 Carole Hooven, *T: The Story of Testosterone, the Hormone that Divides and Dominates Us* (New York: Henry Holt and Company, 2021), 181.

27 For instance, the work of Gary Slutkin is predicated on approaching violence as a contagious disease. While there are some critics of this approach, Slutkin has been impressively successful in helping implement policies and practices that reduce violence. See Gary Slutkin, Charles Ransford, and R. Brent Decker, "Cure Violence: Treating Violence as a Contagious Disease," in *Envisioning Criminology*, eds. Michael D. Maltz and Stephen K. Rice (Switzerland: Springer, 2015), https://doi.org/10.1007/978-3-319-15868-6_5.

28 L. Rowell Huesmann, "The Contagion of Violence," in *The Cambridge Handbook of Violent and Aggressive Behavior*, eds. Alexander T. Vazsonyi, Daniel J. Flannery, and Matt DeLisi (Cambridge, UK: Cambridge University Press, 2018), 527–556.

29 Ibid., 539.

30 Ibid., 540–41. Huesmann describes the narrative of how a young child growing up in such terrible conditions develops his own violent and aggressive responses: "A boy who has grown up observing more . . . violence around him almost every day will believe that the world is a more hostile place than do other boys and will be biased toward making hostile attributions about those who annoy him. He will also have . . . a larger and more accessible repertoire of aggressive scripts over time while observing violence. This will then make

it more likely that he will rehearse the aggressive scripts through fantasizing. His hostile attributions about peers coupled with his readily accessible repertoire of aggressive scripts then make it more likely that he will behave aggressively and violently toward peers. Additionally, from his exposure to violence, he will be more likely to have acquired . . . beliefs that such aggressive and violent behavior is appropriate. Finally, his repeated exposures to violence will have blunted the negative emotional responses (anxiety and fear) that humans normally experience when they see violence or think about violence that might inhibit such behavior. Thus, the more a youth is exposed to war violence, the most certainly he or she will be infected with violence and behave violently toward others."

31 Wilson, *On Human Nature*, 106.

32 Richard A. Lippa, "Sex Differences in Sex Drive, Sociosexuality, and Height Across 53 Nations: Testing Evolutionary and Social Structural Theories," *Archives of Sexual Behavior* 38 (2009), 647. Lippa is speaking here about sexual predispositions, but I think his quote applies to many aspects of human nature.

33 Sarah Hrdy, *Mothers and Others: The Evolutionary Origins of Mutual Understanding* (Cambridge, Mass.: Belknap Press, 2009), 3.

34 World Health Organization, *Violence Against Women Prevalence Estimates, 2018* (Geneva, 2021), XII, available at https://www.who.int/publications/i/item/9789240022256.

35 Richard Wrangham, "Two Types of Aggression in Human Evolution," *Proceedings of the National Academy of Science* 115, No. 2 (Jan. 9, 2018), 248.

36 Wrangham, *Goodness Paradox*. See also Brian Hare and Michael Tomasello, "Human-like Social Skills in Dogs?" *Trends in Cognitive Science* 9, No. 9 (Sept. 2005), 439–444. A fascinating example of domestication is that of the silver fox, which was intentionally domesticated over many generations by a team of Russian scientists led by Dmitri Belyaev and Lyudmila Trut. See Lee Alan Dugatkin and Lyudmila Trut, *How to Tame a Fox (and Build a Dog): Visionary Scientists and a Siberian Tale of Jump-Started Evolution* (Chicago: University of Chicago Press, 2017). Anthropologist Brian Hare and others argue that bonobos (*Pan paniscus*) may have undergone a process of self-domestication, similar to the one proposed for humans. Aside from chimpanzees, bonobos are the most closely related species of primate to humans and they are not as aggressive as chimpanzees. See Brian Hare, Victoria Wobber, and Richard Wrangham, "The Self-Domestication Hypothesis: Evolution of Bonobo Psychology is Due to Selection Against Aggression," *Animal Behaviour* 83, No. 3 (2012), 573–585.

37 Wrangham, *Goodness Paradox.*

38 Boehm, *Moral Origins*, 70.

39 Joan B. Silk, Sarah F. Brosnan, Jennifer Vonk, Joseph Henrich, Daniel J. Povinelli, Amanda S. Richardson, Susan B. Lambeth, Jenny Mascaro, and Steven J. Schapiro, "Chimpanzees Are Indifferent to the Welfare of Unrelated Group Members," *Nature* 437, No. 7063 (2005), 1357–1359.

40 Felix Warneken and Michael Tomasello, "Altruistic Helping in Human Infants and Young Chimpanzees," *Science* 311 (2006), 1301–1303. Felix Warneken and Michael Tomasello, "Helping and Cooperation at 14 Months of Age," *Infancy* 11, No. 3 (2007), 271–294. Felix Warneken, et al., "Spontaneous Altruism by Chimpanzees and Young Children," *PloS Biology* 5 (2007), 1414–1420.

41 Felix Warneken and Michael Tomasello, "Extrinsic Rewards Undermine Altruistic Tendencies in 20-month-olds," *Developmental Psychology*, 44 No. 6 (2008), 1785–1788.

42 Some might argue that there is still some element of culture in the cooperative instincts of one-year-old children. Those infants who come from abusive homes react in distinct ways and show less altruism. Nonetheless, in most normal circumstances, it seems we are pre-wired to cooperate and to be altruistic.

43 Tomasello, *Why We Cooperate*, 7–13.

44 See Karl G. Heider, *Grand Valley Dani: Peaceful Warriors* (Belmont, Calif.: Wadsworth, 1997).

45 Wrangham, *The Goodness Paradox*, 16.

46 Maurice Davie, *The Evolution of War: A Study of Its Role in Early Societies* (New Haven, Conn.: Yale University Press, 1929), 18.

47 As quoted in Davie, *The Evolution of War*, 18.

48 Davie, *The Evolution of War*, 20.

49 Anthropologist Maurice Davie, writing in 1929, summarized the observations from dozens of tribes. He concluded that the overall sentiment between different tribes was one of extreme "isolation, suspicion, [and] hostility. But within a tribe the situation is quite otherwise. Even though there may be [intergroup] conflict at times, the members of a group have a united and common interest against every other group. This is the natural effect of the competition of life. Each group is carrying on a struggle for existence, and its members therefore have interests in common. But in its struggle, it encounters . . . other groups, to which its relation. . . . is generally one of hostility and war. The collision of groups in the struggle for existence, however, tends to bind more closely the members of any one group. Thus, a

distinction arises between one's own tribe and others . . . Different sentiments prevail in regard to the two: between members of the 'we-group,' peace and cooperation exist; their sentiment toward outsiders is that of hatred and hostility." *The Evolution of War*, 16.

50 The work of Polish psychologist Henri Tajfel showed that people could be divided into groups based on the subtlest of distinctions and they would manifest favoritism toward the in-group and discriminate against the out-group. This finding persisted even when people were assigned to be part of different groups randomly and *were told that these groups were random* (i.e., they had no basis in real or perceived similarities among group members). See Henri Tajfel, "Experiments in Intergroup Discrimination," *Scientific American* 233, No. 5 (Nov. 1970), 96–103. See also Michael Billig and Henri Tajfel, "Social Categorization and Similarity in Intergroup Behaviour," *European Journal of Social Psychology* 3, No. 1 (Jan./March 1973), 27–52.

51 Mark Levin, Amy Prosser, David Evans, and Stephen Reicer, "Identity and Emergency Intervention: How Social Group Membership and Inclusiveness of Group Boundaries Shape Helping Behavior," *Personality and Social Psychology Bulletin* 31, No. 4 (April 1, 2005), 443–453.

52 Yarrow Dunham, Andrew Scott Baron, and Susan Carey, "Consequences of Minimal Group Affiliation in Children," *Child Development* 82, No. 3 (May/June 2011), 793–811.

53 Muzafer Sherif, O. J. Harvey, B. Jack White, William R. Hood, and Carolyn W. Sherif., *Intergroup Conflict and Cooperation: The Robbers Cave Experiment* (Norman: University of Oklahoma, 1961). For helpful summaries of this experiment (whose conclusions I also drew from), see Christakis, *Blueprint*, 269–272 and Maria Konnikova, "Revisiting Robbers Cave: The easy spontaneity of intergroup conflict," *Scientific American*, September 5, 2012, https://blogs.scientificamerican.com/literally-psyched/revisiting-the-robbers-cave-the-easy-spontaneity-of-intergroup-conflict/. See also Peter Gray, *Psychology* (New York: Worth Publishers, 2002), 565–566.

54 Muzafer Sherif, et al., *Intergroup Conflict and Cooperation: The Robbers Cave Experiment* (Norman: University of Oklahoma, 1961).

55 Samuel Bowles, "Conflict: Altruism's Midwife," *Nature* 456, (Nov. 20, 2008), 326–327. See also Jung-Kyoo Choi and Samuel Bowles. "The Coevolution of Parochial Altruism and War." *Science* 318 (Oct. 26, 2007), 636–640.

56 Christakis, *Blueprint*, 273.

57 Choi and Bowles, "Coevolution," 640.

58 I first remember hearing this maxim from a colleague, Christopher Pittenger, in the Yale Department of Psychiatry.

59 Muzafer Sherif, et al., *Intergroup Conflict and Cooperation*, 1961.

60 Ibid.

61 Wrangham, *Goodness Paradox*, 123. See also Curtis Marean, "An Evolutionary Anthropological Perspective on Modern Human Origins," *Annual Review of Anthropology* 44 (Oct. 2015), 533–556.

62 Malthus, *An Essay on the Principle of Population*, 1798. Interestingly, the treatise was first published anonymously, but Malthus was soon discovered to be its author. The time period on which Malthus chiefly focused was not one in which our ancestors evolved but rather on his present day of the late 18th century. Nevertheless, his ideas were influential to Darwin, who imagined what life might have been like while our ancestors were evolving.

63 Paul Ehrlich, a biologist at Stanford, has continued Malthus's arguments into more modern times. He published a controversial book in 1968, *The Population Bomb.* Like Malthus, Ehrlich argued that mass starvation would occur due to human overpopulation. In specific, he argued that hundreds of millions of people would starve to death in the 1970s and 1980s, which did not happen (the number of people dying each year from starvation has actually been decreasing over time). More recently, economists Marian Tupy and Gale Pooley have argued compellingly that resources actually keep getting more abundant because of improvements in technology and that a growing population is, on average, a positive thing. See Marian L. Tupy and Gale L. Pooley, *Superabundance: The Story of Population Growth, Innovation, and Human Flourishing on an Infinitely Bountiful Planet* (Washington: Cato Institute, 2022).

64 Jane B. Lancaster and Chet S. Lancaster, "The Watershed: Change in Parental-Investment and Family-Formation Strategies in the Course of Human Evolution," in *Parenting Across the Life Span*, ed. Jane B. Lancaster (New York: Aldine, 1987), 187.

65 Anthropologist Christopher Boehm argues that humans did find another way than simply striving against one another: "That we became efficient, cooperative, equal-opportunity meat sharers was important to our overall evolutionary success. . . . At some level . . . Homo sapiens, with its relatively large social brain . . . understood something about the importance of hunting cooperatively and about the advantages of sharing meat among the entire band. Today, egalitarian hunter-gatherers definitely seem to appreciate the advantages of having a band with more hunters, for obviously this means that the big carcasses to share will come in more often and hence there will be fewer hiatuses when people simply have no major meat to eat." See Boehm, *Moral Origins*, 142.

66 David Rand and his colleagues conducted a series of experiments using social games. They tested whether groups performed better when they were pitted against each other (zero-sum contests) or when they were merely given the same incentive to cooperate without being in direct competition with other groups (Rand's version of a *superordinate goal*). The results showed that competition with another group did foster in-group cooperation. However, providing a noncompetitive incentive for cooperation was just as effective. See Matthew R. Jordan, Jillian J. Jordan, and David G. Rand, "No Unique Effect of Intergroup Competition on Cooperation: Noncompetitive Thresholds Are as Effective as Competitions Between Groups for Increasing Human Cooperative Behavior," *Evolution and Human Behavior* 38 (2017), 102–108.

67 See Tomasello, *Why We Cooperate*, 63–64.

68 Kipling D. Williams and Steve A. Nida, "Ostracism: Consequences and Coping," *Current Directions in Psychology* 20, No. 2 (2011), 71–75.

69 See Boehm, *Moral Origins.*

70 The idea that natural selection can operate on whole groups of distantly related humans was once hotly contested. However, increasing evidence suggests that this may have played some role (and may help explain why we are so much more cooperative than other primates). As Ernst Fehr and Urs Fischbacher state, "In other primate societies, cooperation is orders of magnitude less developed than it is among humans, despite our close, common ancestry." (Increasingly, it is thought that this group selection mechanism may have had more to do with cultural processes than genetic ones.) See Ernst Fehr and Urs Fischbacher, "The Nature of Human Altruism," *Nature* 425 (Oct. 23, 2003), 785. As another example, in modern hunter-gatherer tribes (proxies for what scientists think our ancestors were like), only about 40 percent of people are blood relatives. Thus, there are limits to how much kin selection can explain regarding the extraordinarily cooperative nature of humans. See Kim R. Hill, Robert S. Walker, Miran Božičević, James Eder, Thomas Headland, Barry Hewlett, A. Magdalena Hurtado, Frank Marlow, Polly Wiessner, and Brian Wood, "Co-Residence Patterns in Hunter-Gatherer Societies Show Unique Human Social Structure," *Science* 331, No. 6022 (March 11, 2011), 1286–1289.

Chapter 6: Lust and Love

1 Says Professor Augustin Fuentes: "Our cultural expectation of sexual monogamy is at odds with our evolutionary heritage and basic biology. . . . It is not human nature to seek . . . a . . . sexually monogamous romantic relationship." From *Race, Monogamy, and Other Lies They Told You: Busting*

Myths about Human Nature (Berkeley: University of California Press, 2012), 192.

2 Several cultures throughout history have tolerated polygamy. Almost always, this has been of the form where a man can marry more than one wife (polygyny), as opposed to one woman marrying several men (polyandry). Some scholars believe that monogamy developed in humans some 300,000 years ago, when our ancestors began to organize themselves in hunter-gatherer groups. The thinking is that monogamy dominated human social groups for a few hundred thousand years. However, when humans started settling down (as opposed to being nomadic) some 12,000 years ago, economic inequality began to arise because some individuals were able to accumulate more resources than others. (The accumulation of large quantities of resources wasn't really possible when humans were nomads because these things could not easily be carried around.) It was this rise in inequality that allowed for the wealthiest men to afford and provide materially for multiple wives and their children. The shift back to monogamy has occurred relatively recently (in the last 2,000 years or so) and is still occurring in some areas of the world. So, while the shift from promiscuity to monogamy might not have been linear, humans developed the capacity for both these mating patterns based on their evolutionary history. See Christakis, *Blueprint*.

3 William R. Jankowiak and Edward F. Fischer, "A Cross-Cultural Perspective on Romantic Love," *Ethnology* 31, No. 2 (April 1992), 149–155.

4 Pavel Goldstein, Irit Weissman-Fogel, Guillaume Dumas, and Simone G. Shamay-Tsoory, "Brain-to-Brain Coupling During Handholding is Associated with Pain Reduction," *Proceedings of the National Academy of Science* 115, No. 11 (March 13, 2018).

5 Jarred Younger, Arthur Aron, Sara Parke, Neil Chatterjee, and Sean Mackey, "Viewing Pictures of a Romantic Partner Reduces Experimental Pain: Involvement of Neural Reward Systems," *PloS ONE* 5, No. 10 (Oct. 2010), https://doi.org/10.1371/journal.pone.0013309.

6 David M. Buss, Randy J. Larsen, Drew Westen, and Jennifer Semmelroth, "Sex Differences in Jealousy: Evolution, Physiology, and Psychology," *Psychological Science* 3, No. 4 (July 1992), 251–255.

7 David M. Buss, *The Evolution of Desire: Strategies of Human Mating* (New York: Basic Books, 2016 [originally published in 1994]), 217.

8 See Steven Pinker, *How the Mind Works* (New York and London: W.W. Norton & Company, 1997), 489.

9 David M. Buss, "Jealousy, The Necessary Evil," *Los Angeles Times*, Feb. 14, 2007, https://www.latimes.com/archives/la-xpm-2007-feb-14-oe-buss14-story.html.

10 It's true that other species that are primarily promiscuous may also display violence in response to mate poaching. However, the degree to which sexual jealousy is experienced by both males and females seems to be somewhat unique to humans.

11 Nicholas Christakis makes a great analysis for this argument in Chapter 5 of *Blueprint.*

12 This exercise comes from Robert Sapolsky, *Behave: The Biology of Humans at Our Best and Worst* (New York: Penguin, 2017).

13 It's not exactly clear which came first. A starting point of only having a few kids could select for paternalism; conversely, a starting point of paternalism could select for only having a few kids. In reality, they may have both occurred concurrently over evolutionary time.

14 Sapolsky, *Behave*, 356.

15 See, for example, David P. Barash and Judith Eve Lipton, *The Myth of Monogamy: Fidelity and Infidelity in Animals and People* (New York: W.H. Freeman/Henry Holt and Company, 2002). See also Augustin Fuentes, *Race, Monogamy, and Other Lies They Told You.*

16 D. Lukas and T. H. Clutton-Brock, "Evolution of Social Monogamy," *Science* 341, No. 6145 (Aug. 2, 2013), 526–530.

17 See a similar analysis by Jonathan Haidt, *The Happiness Hypothesis: Finding Modern Truth in Ancient Wisdom* (New York: Basic Books, 2006).

18 Another important driver of enhanced intelligence was likely related to the increasing complexity of social interaction in human groups.

19 David F. Bjorkland, "The Role of Immaturity in Human Development," *Psychological Bulletin* 122, No. 2 (1997), 153–169. See also Zachary Cofran, "Brain Size Growth in Wild and Captive Chimpanzees (Pan troglodytes)," *American Journal of Primatology* 80 (2018), e22876.

20 Bjorklund, "The Role of Immaturity in Human Development."

21 Studies from modern hunter-gatherer societies have suggested that mothers of young children cannot gather enough food for themselves and their offspring without help. See Hillard Kaplan, Kim Hill, Jane Lancaster, and A. Magdalena Hurtado, "A Theory of Human Life History Evolution: Diet, Intelligence, and Longevity," *Evolutionary Anthropology* 9, No. 4 (2000), 156–185.

22 Some scholars have argued that players other than the father were perhaps more important in the development of human offspring. For instance, the scholar Kristin Hawkes has argued that grandmothers were more important to providing sufficient calories for young than fathers. This is the so-called "grandmother hypothesis." See Kristen Hawkes, "Grandmothers and the

Evolution of Human Longevity," *American Journal of Human Biology* 15 (2003), 380–400. While it's likely true that helpers from the extended family could be very beneficial (especially to consuming sufficient calories), it's unlikely that grandmothers were more important than fathers in the overall development of children. My view in this regard is based on a very large sociological literature from studies showing that, at least in modern societies, fatherlessness is a much greater problem for children's development than grandmotherlessness. See, for instance, Popenoe, *Life Without Father*; also, David Blankenhorn, *Fatherless America*.

23 Bjorklund, "The Role of Immaturity in Human Development."

24 David Buss, *Evolutionary Psychology: The New Science of the Mind* (New York: Routledge, Taylor & Francis Group, 2015), fifth ed., 134.

25 R. Chris Fraley, Claudia C. Brumbaugh, and Michael J. Marks, "The Evolution and Function of Adult Attachment: A Comparative and Phylogenetic Analysis," *Journal of Personality and Social Psychology* 89, No. 5 (2005), 731–746.

26 David M. Buss and David P. Schmitt, "Sexual Strategies Theory: An Evolutionary Perspective on Human Mating," *Psychological Review* 100, No. 2 (1993), 204–232. See also Buss, *Evolutionary Psychology*.

27 Buss, *Evolutionary Psychology*.

28 I should note that other scholars have hypothesized that monogamy is a cause, not a consequence, of higher paternal investment. See D. Lukas and T. H. Clutton-Brock, "Evolution of Social Monogamy," *Science* 341, No. 6145 (Aug. 2, 2013), 526–530. Lukas and Clutton-Brock argue that paternal care is not seen in some examples of monogamy in animals, thus it must not have been necessary as a prerequisite to monogamy. However, as they point out and as we have seen from Chapter 2, there is almost always more than one path to the same evolutionary outcome. This is convergent evolution. Monogamous relationships have evolved at least sixty-one different times independently (according to Lukas and Clutton-Brock). Hence, it is possible that in some instances, heavy paternal care in rearing offspring preceded and was a prerequisite to monogamy.

29 John Bowlby, "The Nature of the Child's Tie to His Mother," *International Journal of Psychoanalysis* 39 (1958), 350–373.

30 R. Chris Fraley, Claudia C. Brumbaugh, and Michael J. Marks, "The Evolution and Function of Adult Attachment: A Comparative and Phylogenetic Analysis."

31 Cindy Hazan and Phillip Shaver, "Romantic Love Conceptualized as an Attachment Process," *Journal of Personality and Social Psychology* 52, No. 3 (1987), 511–524.

32 See, for example, Dirk Scheele, Andrea Wille, Keith M. Kendrick, Birgit Stoffel-Wagner, Benjamin Becker, Onur Güntürkün, Wolfgang Maier, and René Hurlemann, "Oxytocin Enhances Brain Reward System Responses in Men Viewing the Face of Their Female Partner," *Proceedings of the National Academy of Sciences* 110, No. 50 (December 2013): 20308–20313.

33 William R. Jankowiak and Edward F. Fischer, "A Cross-Cultural Perspective on Romantic Love," *Ethnology* 31, No. 2 (April 1992), 154.

34 Buss, *Evolutionary Psychology*.

35 This work was done by Peggy La Cerra as part of a doctoral dissertation, as described in Buss, *Evolutionary Psychology* (Chapter 4). Confirmation of these findings has been found by Gary L. Brase, "Cues of Parental Investment as a Factor in Attractiveness," *Evolution and Human Behavior* 27, No. 2 (March 2006), 145–157. In these studies, women consistently report attraction to cues that a man is willing to invest in childcare. When the roles are reversed, such cues have almost no effect on men's preference for women.

36 James R. Roney, Katherine N. Hanson, Kristina M. Durante, and Dario Maestripieri, "Reading Men's Faces: Women's Mate Attractiveness Judgments Track Men's Testosterone and Interest in Infants," *Proceedings of the Royal Society B* 273 (2006), 2169–2175.

37 The colorful plumage that male peacocks have evolved may be a signal of superior health. Adeline Loyau, Michel Saint Jalme, Cécile Cagniant, and Gabriele Sorci, "Multiple Sexual Advertisements Honestly Reflect Health Status in Peacocks (*Pavo cristatus*)," *Behavioral Ecology and Sociobiology* 58 (May 3, 2005), 552–557.

38 Similar to the peacock example, one of the reasons that female guppies may prefer bright colors is it may be a sign of better health. Guppies with drab coloration are often afflicted with parasites. See Jean-Guy J. Godin and Heather E. McDonough, "Predator Preference for Brightly Colored Males in the Guppy: A Viability Cost for a Sexually Selected Trait," *Behavioral Ecology* 14, No. 2 (March 2003), 194–200.

39 I should point out that natural selection, as I've argued in earlier chapters, doesn't always push in the direction of selfishness. For arguments about how female sexual selection may have contributed to altruism, see Steven Arnocky, Tina Piche, Graham Albert, Danielle Oulette, and Pat Barclay, "Altruism Predicts Mating Success in Humans," *British Journal of Psychology* 108, No. 2 (May 2017), 416–435.

40 Some biologists will take issue with this and say that females are driven by mate preferences that will enhance the genetic fitness of their children. This would, in some circumstances, lead women to choose men who are not

necessarily prone to long-term commitment. There is undoubtedly some truth to this, which is another example of how different evolutionary pressures pushed in different directions, creating conflicting dispositions with respect to human mating preferences.

41 David Sloan Wilson gives another interesting example of how female mate preference can affect male mating behavior. Citing work from one of his doctoral students, he describes the mating patterns of water striders. Male water striders can have two very different approaches to mating. Some male water striders act very aggressively, trying to mate with as many females as possible, regardless of whether the female shows any signs of receptivity. We'll call this the aggressive strategy. Other male water striders mate only when enticed by females. This would be the gentleman strategy. You would think that, among the two approaches, the aggressive strategy would win out in an evolutionary contest. But it turns out, the aggressive strategy is so disturbing to the females that they become stressed out, eat less, and lay fewer eggs. Plus, the females tend to gravitate toward and prefer the males who are more gentlemanly in their courting approaches. Hence, natural populations of water striders can contain both approaches to mating in part because of female mate preference. See David Sloan Wilson, *Does Altruism Exist?* (New Haven, Conn., and London: Yale University Press, 2015), 24.

42 Technically, it's slightly skewed toward females because women generally live about 5 percent longer than men, so at any given time, there will generally be more women alive than men.

43 A. Kosztolányi, Z. Barta, C. Küpper, and T. Székely, "Persistence of an Extreme Male-Biased Adult Sex Ratio in a Natural Population of Polyandrous Bird," *Journal of Evolutionary Biology* 24, No. 8 (Aug. 2011), 1842–1846.

44 Joseph Henrich, Robert Boyd, and Peter J. Richerson, "The Puzzle of Monogamous Marriage," *Philosophical Transactions of the Royal Society B* 367 (2012), 657–669.

45 As noted above (see endnote 2 for this chapter), many scholars believe that monogamy developed in humans about 300,000 years ago (from a more promiscuous past before that), when our ancestors began to organize themselves in hunter-gatherer groups. It is believed there was a return to polygamous groups (at least for some groups) when humans started settling down (as opposed to being nomadic) and economic inequality arose.

46 See Russell D. Clark and Elain Hatfield, "Gender Differences in Receptivity to Sexual Offers," *Journal of Psychology & Human Sexuality* 2, No. 1 (1989), 39–55.

Chapter 7: Free to Choose

1 Sam Harris, *Free Will* (New York: Free Press, 2012), 1.

2 The United States Supreme Court has explicitly stated that "belief in freedom of the human will" is foundational to the legal system, calling it a "universal and persistent" premise, distinct from "a deterministic view of human conduct that is inconsistent with the underlying precepts of our criminal justice system." See *United States v. Grayson,* 438 U.S. 41 (1978).

3 David Wisniewski, Robert Deutschländer, and John-Dylan Haynes, "Free Will Beliefs Are Better Predicted by Dualism Than Determinism Beliefs Across Different Cultures," *PLOS ONE* 14, No. 9 (Sept. 2019): e0221617.

4 Tamar Kushnir, Alison Gopnik, Nadia Chernyak, Elizabeth Seiver, and Henry M. Wellman, "Developing Intuitions About Free Will Between Ages Four and Six," *Cognition* 138 (2015), 79–101.

5 This definition of free will can be traced as early as the iconic psychologist and philosopher William James (1842–1910). See Bob Doyle, "Jamesian Free Will: The Two-Stage Model of William James," *William James Studies* 5 (2010), 1–28.

6 See, for instance, Christopher Taylor and Daniel Dennett, "Who's Afraid of Determinism? Rethinking Causes and Possibilities," in *The Oxford Handbook of Free Will*, ed. Robert Kane (Oxford, UK: Oxford University Press, 2005), 257–277.

7 For many of the arguments of this chapter, including language describing alternative possibilities and causal mental control, I lean heavily on Christian List's impressively readable book, *Why Free Will Is Real* (Cambridge, Mass.: Harvard University Press, 2019). List includes a third requirement for free will: intentional agency.

8 Again, here I borrow ideas and language from List, *Why Free Will Is Real*, 2–3.

9 Quote from Shaun Nichols, "Experimental Philosophy and the Problem of Free Will," *Science* 331, No. 6023 (March 18, 2011), 1401.

10 See List, *Why Free Will Is Real*, 41.

11 Some purist philosophers might assert that science does not have claim to the question of whether a given system is deterministic or not. (For instance, some believe that quantum mechanics might yet be deterministic because of a so-called hidden variable that we have not yet discovered.) These people might argue that science can only describe what appears to be determinism or indeterminism. If we follow this same logic, then it could be concluded that—if free will hinges on some level of indeterminism (as I believe)—then science cannot prove or disprove the existence of free will.

12 This argument is adapted from List, *Why Free Will Is Real.*

13 P. W. Anderson, "More Is Different: Broken Symmetry and the Nature of the Hierarchical Structure of Science," *Science* 177, No. 4047 (Aug. 4, 1972), 393–396.

14 Of course, inflation has to do with how much money is being circulated, so (like all analogies) if we take this too far it breaks down. (You might be able to deduce that something akin to inflation is occurring if the government suddenly prints much more money.) Another example is the concept of *equivalence*. In the United States, people value four quarters the same as one dollar bill. Yet it would be impossible to deduce from the physical examination of the money what makes them equal in value. But the concept of equivalence is real and critical to the principles governing economics.

15 Elizabeth Barnes, "Emergence and Fundamentality," *Mind* 121, No. 484 (Oct. 2012), 873–874.

16 Bryson, *The Body*, 4.

17 Philosopher Eddy Nahmias would go even further and argue that all levels are indeterministic even if only one level is indeterministic. For instance, even if only the quantum physics level is indeterministic, this means that the scientific measurements scientists make in the macroscopic world to track the uncertainty of subatomic particles reflect the indeterminism of the quantum level. Personal interview, December 13, 2021.

18 See List, *Why Free Will Is Real.*

19 Jeremy Butterfield, "Laws, Causation and Dynamics at Different Levels," *Interface Focus* 2 (Feb. 2012), 108.

20 See Paul W. Glimcher, "Indeterminacy in Brain and Behavior," *Annual Review of Psychology* 56 (2005), 25–56.

21 Björn Brembs, "The Brain as a Dynamically Active Organ," *Biochemical and Biophysical Research Communications* 564 (July 30, 2021), 55–69.

22 See, for instance, K. L. Briggman, H.D.I. Abarbanel, and W. B. Kristan Jr., "Optimal Imaging of Neuronal Populations During Decision-Making," *Science* 307, No. 5711 (Feb. 11, 2005), 896–901.

23 Alexander Maye, Chih-hao Hsieh, George Sugihara, and Björn Brembs, "Order in Spontaneous Behavior," *PLoS ONE* 5 (May 16, 2007). Maye and his colleagues have broken down the behavior of the fruit fly into different components. Some of the fly's behavior is driven by inputs from the environment, such as smells, sights, or sounds that are encountered. But there is another part of fly behavior that doesn't seem to rely on these inputs. For a long time, many neuroscientists assumed that this "residual" part of behavioral variability was just "noise" from the environment. But Brembs and his team have discovered that there is an element of unpredictability and

spontaneity that comes from the fly itself. The behavior of the fly cannot merely be reduced to responses to the outside world: "Even fly brains are more than just input/output systems" [Maye, et al., e443].

24 Many philosophers would argue that whether the world is deterministic or not is more of a metaphysical claim than something that can or cannot be proved by science. But on a practical level, it appears that humans (as well as other organisms) can and do behave in ways that cannot be distinguished from indeterministic behavior.

25 This particular version of the Harvard Law of Animal Behavior comes from Joel Garreau's book, *Radical Evolution* (New York: Broadway Books, 2005), 153. Despite the half-serious tone of the Harvard Law, some biologists are starting to recognize the importance of its main tenet. As the biologist Paul Grobstein states: "The Harvard Law in fact has quite concrete and deep significance for understanding the basic information processing characteristics which underlie the behavior of all organisms, humans very much included. An appreciation of this requires drawing together threads from a variety of lines of inquiry and is facilitated by a perspective that treats both behavior and the nervous system as nested sets of interacting information-processing boxes each with more or less clearly defined inputs and outputs. Briefly put, the Harvard Law provides the basis for a desirable and productive fusion of scientific and folk perspectives on the determinants of behavior, one which acknowledges that some degree of unpredictability is not only inevitable but desirable." See Paul Grobstein, "Variability in Brain Function and Behavior," in *The Encyclopedia of Human Behavior, Volume 4*, ed. V. S. Ramachandran (Cambridge, Mass.: Academic Press, 1994), 448.

26 Martin Heisenberg, "Is Free Will an Illusion?" *Nature* 459 (May 14, 2009), 164–165. At another point in this essay, Heisenberg argues: "Almost 100 years ago, quantum physics eliminated a major obstacle to our understanding of [free will] when it disposed of the idea of a Universe determined in every detail from the outset. It uncovered an inherent unpredictability in nature, in that we can never know precisely at a given moment all properties of a particle—such as both its position and its momentum" (164). Coincidentally, Martin is the son of Werner Heisenberg, who developed this theory, the Heisenberg Uncertainty Principle.

27 Björn Brembs, "Towards a Scientific Concept of Free Will as a Biological Trait: Spontaneous Actions and Decision-Making in Invertebrates," *Proceedings of the Royal Society* B 278 (2011), 930–939.

28 These examples might lead to the question of whether animals have some sense of free will. They indeed might. For an interesting discussion on this

issue, see Chapter 6 of *Mama's Last Hug*, by Frans de Waal (New York: W.W. Norton & Co., 2019).

29 Glimcher, "Indeterminacy in Brain and Behavior," 52.

30 This concept is sometimes referred to as Hume's fork (attributed to philosopher David Hume): "Either our actions are determined, in which case we are not responsible for them, or they are the result of random events, in which case we are not responsible for them."

31 Robert Trattner Sherman and Craig A. Anderson, "Decreasing Premature Termination from Psychotherapy," *Journal of Social and Clinical Psychology* 5, No. 3 (1987), 298–312.

32 Robert L. Woolfolk, Mark W. Parrish, and Shane M. Murphy, "The Effects of Positive and Negative Imagery on Motor Skill Performance," *Cognitive Therapy and Research* 9, No. 3 (1985), 335–341.

33 See Stephen M. Kosslyn and Samuel T. Moulton, "Mental Imagery and Implicit Memory," in *The Handbook of Imagination and Mental Simulation*, eds. Keith D. Markman, William M. P. Klein, and Julie A. Suhr (New York: Psychology Press, 2009), 35–51.

34 Anne M. Theiler and Louis G. Lippman, "Effects of Mental Practice and Modeling on Guitar and Vocal Performance," *Journal of General Psychology* 122, No. 4 (1999), 329–343.

35 Dirk C. Prather, "Prompted Mental Practice as a Flight Simulator," *Journal of Applied Psychology* 57, No. 3 (1973), 353–55.

36 Charles W. Sanders, Mark Sadoski, Richard M. Wasserman, Robert Wiprud, Mark English, and Rachel Bramson, "Comparing the Effects of Physical Practice and Mental Imagery Rehearsal on Learning Basic Venipucture by Medical Students," *Imagination, Cognition, and Personality* 27, No. 2 (2007), 117–127.

37 Sarah Milne, Sheina Orbell, and Paschal Sheeran, "Combining Motivational and Volitional Interventions to Promote Exercise Participation: Protection Motivation Theory and Implementation Intentions," *British Journal of Health Psychology* 7 (2002), 163–184.

38 Sheina Orbell, Sarah Hodgkins, and Paschal Sheeran, "Implementation Intentions and the Theory of Planned Behavior," *Personality and Social Psychology Bulletin* 23, No. 9 (1997), 945–954.

39 Vered Murgraff, David White, and Keith Phillips, "Moderating Binge Drinking: It is Possible to Change Behaviour If You Plan It in Advance," *Alcohol and Alcoholism* 31, No. 6 (1996), 577–582.

40 Peter M. Gollwitzer and Paschal Sheeran, "Implementation Intentions and Goal Achievement: A Meta-analysis of Effects and Processes," *Advances in*

Experimental Social Psychology 38 (2006), 69–119. For a helpful review of many of these examples with less math, see Peter Gollwitzer, "Implementation Intentions: Strong Effects of Simple Plans," *American Psychologist* 54, No. 7 (July 1999), 493–503.

41 Roy F. Baumeister, E. J. Masicampo, and Kathleen D. Vohs, "Do Conscious Thoughts Cause Behavior?" *Annual Review of Psychology* 62 (2011), 331–361.

42 Ibid., 331. It is interesting to note that there is a much larger body of research that supports the notion that our conscious thoughts control behavior than the small handful of studies that vocal opponents of free will have propped up as evidence that free will does not exist (such as the Libet experiments, discussed later).

43 It is worth pointing out that in many cases, our unconscious or automatic intentions begin in our consciousness. As we learn a behavior, it can be shifted into the system of unconsciousness. This is how habits form. For instance, when I first moved to my current residence, I would consciously think as I drove to my job. I would have to pay attention as I took the right exit, made the proper turn, and so forth. This is because the area was unfamiliar. But as I performed this action of driving to work every day over and over, it became second nature. Now I no longer consciously think about where I have to turn and what exit I take to arrive at work. In fact, some days I become preoccupied in my thoughts while driving. Before I realize it, I have arrived. Yet for the life of me I cannot remember taking a given street or turning into the parking garage. The part of my brain that was driving was on autopilot. Most people can relate to an experience such as this. Hence, in many cases, our unconscious thoughts and intentions were once in our conscious control. But if there is a change in circumstances, conscious thoughts and intentions can override unconscious ones. If on my drive to work I encounter a detour or a traffic accident, my conscious brain suddenly kicks back into gear and takes control from the autopilot mode.

44 See Stephen Cave, "There's No Such Thing as Free Will," *The Atlantic*, June 2016, https://www.theatlantic.com/magazine/archive/2016/06/theres-no-such-thing-as-free-will/480750/.

45 Benjamin Libet, Curtis A. Gleason, Elwood W. Wright, and Dennis K. Pearl, "Time of Conscious Intention to Act in Relation to Onset of Cerebral Activity (Readiness Potential): The Unconscious Initiation of a Freely Voluntary Act," *Brain* 106 (1983), 623–642.

46 See, for instance, Patrick Haggard and Martin Eimer, "On the Relation Between Brain Potentials and the Awareness of Voluntary Movements." *Experimental Brain Research* 126 (1999), 128–136; I. Keller and H.

Heckhausen, "Readiness Potentials Preceding Spontaneous Motor Acts: Voluntary vs. Involuntary Control," *Electroencephalography and Clinical Neurophysiology* 76 (1990), 351–361.

47 See List, *Why Free Will Is Real.*

48 Eddy Nahmias, "Scientific Challenges to Free Will," in *A Companion to the Philosophy of Action*, eds. Timothy O'Connor and Constantine Sandis (Hoboken, N.J.: Wiley-Blackwell, 2010), 352.

49 Christian List offers the analogy of a committee that is tasked with making a decision on whether to hire someone. There will be a good amount of "lower level" activity among the individual members of the committee leading up to the decision. There will be reading of relevant material (the applicant's resume), discussion of subject matter (how well does the applicant fit the job description), and even interviewing of the candidate (how does she blend in with the company culture). Someone studying the committee decision-making process could liken this activity to a "readiness potential" or an increase in brain activity that precedes the actual decision. But it isn't until the committee officially meets and votes that the decision is final. The fact that preparation for the official vote has occurred does not somehow undermine the committee's decision-making ability. As we learn more and more about the committee, we may even be able to make predictions with respect to the eventual decision. These predictions would be based on how the discussion is going, which way some committee members seem to be leaning, how they have voted in the past, and so forth. But the fact that we can predict the decision based on which way the committee is leaning does not in any way take away from the decision-making power of the committee. See List, *Why Free Will Is Real.* This analogy also helps to contextualize subsequent experiments by John-Dylan Haynes and colleagues. Haynes and his associates devised an experiment thus: brain scans were taken while participants were given tasks that involved making simple decisions. For instance, participants were given two numbers and were told to add them or subtract them. Or they were given the choice to press a button with their right hand or their left (the choice was up to them). In the experiment, researchers devised a *neural signature*, or pattern of brain activity based on brain scans, to attempt to predict whether participants would "choose" whether to add or subtract the numbers they were given. In this particular experiment, Haynes and his colleagues report they could predict the choice of the participants several seconds before they actually reported making the choice. But, and this is an important point, the predictions were far from perfect. In a paradigm with only two choices, their algorithm correctly predicted roughly 60 percent of

choices, which is only 10 percent better than if we were guessing at random. See Chun Siong Soon, Marcel Brass, Hans-Jochen Heinze, and John-Dylan Haynes, "Unconscious Determinants of Free Decisions in the Human Brain," *Nature Neuroscience* 11 (2008), 543–545. Yet anyone can think of several situations where this would not apply. We can make decisions, even conscious ones, much quicker than the 7–10 seconds that Haynes claims sets in deterministic stone our future decisions. As Eddy Nahmias states: "Brain activity cannot, in general, *settle* our choices [several] seconds before we act, because we can react to changes in our situations in less time than that. If we could not, we would all have died in car crashes by now!" See Eddy Nahmias, "Why We Have Free Will," *Scientific American*, Jan. 1, 2015, 78. See also Aaron Schurger, Pengbo "Ben" Hu, Joanna Pak, and Adina L. Roskies, "What Is the Readiness Potential?" *Trends in Cognitive Science* 25, No. 7 (July 2021): 558–570.

50 As Alfred Mele says: "How similar is the arbitrary picking of a button to a decision to ask one's spouse for a divorce—or to change careers or start a small business—after protracted reflection on reasons for and against the decision? If arbitrary picking is not very similar to these other decisions, claiming that what happens in instances of arbitrary picking also happens in instances of complicated, painstaking decision making is a huge stretch." From Alfred Mele, *Free: Why Science Hasn't Disproved Free Will* (Oxford, UK: Oxford University Press, 2014), 30.

51 Nahmias, "Why We Have Free Will," 78–79.

52 In evaluating the choices to which we attach real meaning and importance, the analogy of Christian List's committee process (see above endnote 49 from this chapter) is a much better description than Libet's brain experiments where the decision is made in an instant without prior conscious thought. For further discussion of implications of Libet-type experiments, see Nahmias, "Why We Have Free Will" and Mele, *Free: Why Science Hasn't Disproved Free Will.*

53 This is sometimes referred to as the social-scientific indispensability argument. See List, *Why Free Will Is Real.*

54 Other arguments against free will come from the late psychologist Daniel Wegner. See his book, *The Illusion of Conscious Will* (Cambridge, Mass.: MIT Press, 2003). Wegner argued that, in some instances, we do not have as much conscious control over things as we think we do. Wegner would point out some fascinating but unusual examples to illustrate his point. For instance, in one study, participants were asked to move the index finger of either their left or right hand upon hearing a click. At this point, a magnetic

field was applied to their brain (via a technique called transcranial magnetic stimulation) that stimulated the part of the brain that controls movement. The external stimulation was applied in such a way that the external magnetic field, not the participants, was the source of the index finger moving. In many instances, the research subjects were tricked, so to speak, into believing that they themselves had consciously willed the finger to move. They did not recognize that the magnetic field was the actual causal force. In this instance, the subjects erroneously attributed causality to their own sense of free will when an external source (the magnetic field) was actually the driving force. (This experiment is given in Joaquim P. Brasil-Neto, Alvaro Pascual-Leone, Josep Valls-Sole, Leonardo G. Cohen, and Mark Hallett, "Focal Transcranial Magnetic Stimulation and Response Bias in a Forced-Choice Task," *Journal of Neurology, Neurosurgery, and Psychiatry* 55 (1992), 964–966. Wegner also frequently cites examples of the opposite phenomenon, where individuals erroneously attribute causality to external sources that are clearly being driven by themselves. Such examples include trance channeling, spirit possession, Ouija boards, and dissociative identity disorder. These examples, Wegner argues, show how we can be tricked into misattributing the source of causality, either to ourselves or to other external sources. Finally, Wegner concludes, because of these examples that show how easily we misattribute causality, there is no such thing as free will. But his reasoning is faulty. The fact that the experience of conscious causal control is, in unusual circumstances, sometimes wrong does not mean it is always wrong. We may be more constrained by biological or environmental factors than we would like to admit, but this does not mean that we have no free will, no alternative possibilities, no conscious input to our fate. The best interpretation of the examples that Wegner and others cite is that these illusions of will might be analogous to illusions of the visual system. The fact that visual illusions or even hallucinations occur in rare circumstances does not mean that the visual system is wholly unreliable. Indeed, our fine-tuned visual system is a remarkable accomplishment of our evolution that, for the most part for most people, serves us very well. The fact that our intuitions about our will are occasionally wrong does not mean that they are never right. One further point is worth making, which is that the examples that Wegner uses to make his point are almost always very unusual. In addition to the aforementioned instances (using a magnetic field external to the brain to manipulate body movement), other often-cited examples include various forms of Victorian spiritualism (such as the Ouija boards and trance channeling just mentioned), individuals with split brains (meaning that the connections between the right and left cerebral hemispheres were either surgically severed or never formed), and hypnosis. If we take Wegner at his word, we also might be tempted to conclude that the existence of illness in *someone* rules out the possibility of health in *anyone*. These arguments are adapted from Eddy Nahmias, "Is Free Will an Illusion? Confronting Challenges from the Modern Mind Sciences," in *Moral Psychology, Vol. 4: Freedom and Responsibility*, ed. Walter Sinnott-Armstrong (Cambridge, Mass.: MIT Press, 2014), 1–57. See also Alfred Mele, *Free: Why Science Hasn't Disproved Free Will* (Oxford, UK: Oxford University Press, 2014).

55 As quoted in Viktor Frankl, *Man's Search for Meaning* (New York: Simon and Schuster, 1984 [originally published in 1959]), 153–154.

56 Ibid., 74–75.

57 Ibid., 75, 132.

Chapter 8: The Meaninglessness of Existence Revisited: An Enlightened View of Human Nature

1 Richard Wrangham, *The Goodness Paradox* (New York: Pantheon Books, 2019), 4.

2 Wilson, *Sociobiology*, 547, 551.

3 "Why You Do What You Do," *Time*, Aug. 1, 1977, available at https://content.time.com/time/subscriber/article/0,33009,915181-1,00.html.

4 Helen Fisher, "'Wilson,' They Said, 'You're All Wet!'" *New York Times*, Oct. 16, 1994, https://www.nytimes.com/1994/10/16/books/wilson-they-said-your-all-wet.html?.

5 René Dubos, *So Human an Animal* (New York: Charles Scribner's Sons, 1968), 127–128.

6 Dubos, *So Human an Animal*, 133–134.

7 For example, see writings from Jerry Coyne and Robert Sapolsky.

8 Wilson, *On Human Nature*, 1.

9 It's inevitable in a deep discussion of free will, that someone will ask if animals have free will. Using the framework I've outlined for free will, it has certainly been shown that behavior in many animals is not deterministic (the "free" part of free will). But the kinds of psychological experiments that show how conscious thought influences our behavior (the "will" part) is not really open to experimentation on animals at this point in time. These experiments require sophisticated communication that, up to this point, is only possible through human language. So whether animals have free will or not in the same sense that we seem to have it is, at least from my view, unknown.

10 It should be noted that the operation of natural selection on a higher level than the individual does not always produce admirable qualities. Tribalism is

an epiphenomenon of selfishness acting on a group level. While strong loyalty to kin and groups is important, this loyalty, when taken too far, sometimes breeds antagonism toward other groups. Self-righteousness and retributive moral instinct to punish those who do not conform to social norms are also a result of natural selection (or even cultural evolution) on a level other than the individual. These tendencies can be useful in social groups yet when taken to the extreme can breed animosity and even violence.

11 Of course, there are many mechanisms that created these opposing characteristics. Direct and indirect reciprocity likely played a role. The long period of maturation of the human infant is likely responsible for the comparatively large capacity for compassion from human parents. Sexual selection may have also played a role (see Chapter 6).

12 Again, there are likely many mechanisms that have created these opposing characteristics as well, including direct reciprocity (also known as reciprocal altruism) and indirect reciprocity altruism. Richard Wrangham makes a case for the self-domestication of humans leading to our comparatively low levels of hot violence. Sarah Hrdy argues that the way humans raise their young cooperatively, what is called alloparenting, has implications for the prosocial aspects of human nature. However, it would seem so much of the deepest forms of the prosocial aspects of our nature—what the humanities would label as human virtue—come from the way evolution has shaped our relationships with kin.

13 It's easy to make an evolutionary argument for how a measure of free will—or at least behavioral indeterminism—might have evolved. As an evolving organism, you want to have some measure of unpredictability so that it's difficult for a predator to capture you. But you don't want your behavior to be so totally unpredictable that you thrust yourself into situations that are just as hazardous as being caught by the predator (like running off the face of a cliff). To be clear, the type of indeterminism that neuroscientists such as Bjorn Brembs and his colleagues talk about is not exactly that same level as what many people think of when they conceptualize free will. Nevertheless, it provides a basis for a higher level, moral version of the concept. See Brembs, "The Brain as a Dynamically Active Organ," 2021.

14 As quoted in Dubos, *So Human an Animal,* 226.

15 Robert Louis Stevenson, *The Strange Case of Dr. Jekyll and Mr. Hyde* (Aurora, Colo.: Chump Change, 2017, [original edition 1886]), 9.

16 Richard Wrangham theorizes that Stevenson may have been influenced by the scientific debates of the day. Given that scientists at the time were arguing that man had descended from apes, Stevenson may have conjured that man had some vestige of apelike qualities within him. If we evolved from apes, we

surely harbored somewhere in the dark recesses of our brains a more primitive nature. See Wrangham, *Goodness Paradox*, 222.

17 The Harvard anthropologist Richard Wrangham describes this dual potential of human nature as follows: "The potential for good and evil occurs in every individual. Our biology determines the contradictory aspects of our personalities, and society modifies both tendencies. Our goodness can be intensified or corrupted, just as our selfishness can be exaggerated or reduced. . . . The combination that makes humans odd is that we are both intensely calm in our normal social interactions, and yet in some circumstances so aggressive that we readily kill." See Wrangham, *The Goodness Paradox*, 6–7.

18 Sam Harris, "The Strange Case of Francis Collins," *Sam Harris Blog*, Aug. 6, 2009, https://samharris.org/the-strange-case-of-francis-collins/.

19 Sebastian Herrera, "With Free Haircuts, Houstonian Changes One Disadvantaged Life At a Time," *Houston Chronicle*, Nov. 22, 2017, https://www.houstonchronicle.com/about/houston-gives/article/With-free-haircuts-Houstonian-changes-one-12374650.php.

20 Greg Grienzi, "A Remarkable Partnership Sets High-Schoolers on New Path," *The JHU Gazette*, June 11, 2007, https://pages.jh.edu/~gazette/2007/11jun07/11mentor.html.

21 Moralistic aggression likely developed as a useful emotion to keep free riders from taking advantage of the group.

22 Damien McElroy, "Amish Killer's Widow Thanks Families of Victims for Forgiveness," *Daily Telegraph*, October 16, 2006, https://www.telegraph.co.uk/news/worldnews/1531570/Amish-killers-widow-thanks-families-of-victims-for-forgiveness.html.

23 Richard Sandomir, "Victoria Ruvolo, Who Forgave Her Attacker, Is Dead at 59," *New York Times*, March 28, 2019, available at https://www.nytimes.com/2019/03/28/obituaries/victoria-ruvolo-dead.html.

24 Jim Dwyer, "The Doctor Came to Save Lives. The Co-op Board Told Him to Get Lost," *New York Times*, April 3, 2020, available at https://www.nytimes.com/2020/04/03/nyregion/co-op-board-coronavirus-nyc.html.

25 From a scientific standpoint, we may never know at what point humans developed free will. Perhaps it happened through a long evolutionary process over thousands, if not hundreds of thousands of years. Perhaps it happened relatively quickly from an evolutionary standpoint. Perhaps even in a single generation. I don't presume to know. But when we do find this out, we shouldn't be surprised to recognize that it happened through natural means that at some point we will be able to understand.

26 An important question that would naturally arise is, how is it possible from a mechanistic standpoint to be both selfish and altruistic? The answer likely lies in the human brain, which is amazingly complex but also plastic (changeable). As reviewed before, genes do not code for traits or behaviors. They code for proteins. These proteins in turn have built and maintained the marvelous circuits and networks of the human brain. It is very possible that there is a set of microcircuits that is associated with a given trait such as selfishness. But these circuits are not static. They can change. Through practice—repeated thoughts and behaviors and habits—some of these circuits become more engrained (through what's called long-term potentiation) and become stronger and more automatic. Through disuse, some may atrophy and become very weak, perhaps even nonexistent after an extended period.

27 Wilson, *On Human Nature*, 150.

28 Not too long ago, some ten or twenty years perhaps, the psychological literature seemed to abound with reports that actions and behaviors most people intuitively believed were under our conscious control were in fact driven by the unconscious part of our mind. However, many of these original studies have not been replicated, suggesting that the original findings were not as accurate as originally understood, or perhaps their explanations were not complete. In one famous experiment, college students were asked to solve a scrambled-sentence task that was allegedly part of a language proficiency experiment. Half of the participants worked on scrambled sentences that contained elderly stereotype words, such as retirement, Florida, old, gray, bingo, forgetful, etc. The other half did not have such words. Once the participants finished the task, he or she was thanked for participating and told where the exit was. But the real objective of the study came as the participants walked to the elevator. A study experimenter, posing as a student waiting outside a professor's office, timed the walking pace of each participant. Those in the group who were exposed to elderly stereotyped words walked at a slower pace compared to the control group. See John Bargh, Mark Chen, and Lara Burrows, "Automaticity of Social Behavior: Direct Effects of Trait Construct and Stereotype Activation on Action," *Journal of Personality and Social Psychology* 71, No. 2 (1996), 230–244. However, this study has failed to replicate, and it appears that the original result might have been due more to the expectations of finding an effect on the part of the experimenters. See Stéphane Doyen, Olivier Klein, Cora-Lise Pichon, and Axel Cleeremans, "Behavioral Priming: It's All in the Mind, but Whose Mind?" *PLOS ONE* 7, No. 1 (Jan. 2012), https://doi .org/10.1371/journal.pone.0029081. Nowadays, it seems like psychology has given a little bit more control back to our conscious selves.

29 Jonathan Haidt, *The Happiness Hypothesis: Finding Modern Truth in Ancient Wisdom* (New York: Basic Books, 2006), 4. In Haidt's metaphor, the distinction within us of the rider and the elephant is analogous to the distinction between different systems of thinking (System 1 [automatic, elephant] and System 2 [deliberative, rider]) described by Daniel Kahneman in *Thinking, Fast and Slow* (New York: Farrar, Straus and Giroux, 2011).

Chapter 9: The Good Life

1 George Vaillant, *Adaptation to Life* (Cambridge, Mass.: Harvard University Press, 1977), 3.

2 The two study cohorts were originally called the Grant study (Bock's original group of Harvard students) and the Glueck study (the group of inner-city young men).

3 The electroencephalogram (EEG) tests ultimately performed on the subjects yielded little useful information.

4 Vaillant, *Adaptation to Life*, 315.

5 Ibid., 306–308.

6 Joshua Wolf Shenk, "What Makes Us Happy," *The Atlantic*, June 2009, https://www.theatlantic.com/magazine/archive/2009/06/what-makes-us-happy/307439/.

7 Robert Waldinger, "What Makes a Good Life? Lessons from the Longest Study on Happiness," filmed Nov. 2015, TED Video, https://www.ted.com/talks/robert_waldinger_what_makes_a_good_life_lessons_from_the_longest_study_on_happiness?language=en.

8 Bowlby recognized a grave error from Freud's influential work, *Interpretation of Dreams*. Freud claimed that "when people are absent, children do not miss them with any great intensity; many mothers have learnt [this] to their sorrow." Sigmund Freud, *The Interpretation of Dreams*, transl. James Strachey (New York: Basic Books, 1955), 273. It's hard to understand how this statement went unchallenged for so long. Anyone who has raised a child immediately recognizes the fallacy of this statement in young children.

9 Robert Hinde, who was a mentor to the famous primatologist Jane Goodall, was one of the ethologists that greatly influenced Bowlby.

10 For example, a series of famous but controversial experiments conducted by Harry Harlow demonstrated that social needs can drive behavior. In the 1950s, Harlow observed how monkeys would develop when isolated from their mothers. He separated young monkeys from their mothers and raised them in a cage with two possible surrogate mother figures constructed of wire mesh. At the top of the wire body was something resembling a face. One

wire-mesh mother was equipped with a feeding milk bottle at its midsection. The other had no food to offer but was covered in a soft cloth to provide emotional comfort through cuddling. The monkeys only visited the milk-giving wire-mother when they were hungry. However, they spent significant time with the furry mother. A famous picture, used in many basic psychology textbooks, shows a baby monkey reaching over to drink milk from the wire monkey while still clinging with its hind legs to the comfort of the soft fur of his "real" mother. See Harry F. Harlow, "The Nature of Love," *American Psychologist* 13 (1958), 673–685.

11 John Bowlby, "The Nature of the Child's Tie to His Mother," *The International Journal of Psychoanalysis* 39 (1958), quote on 358 (350–373).

12 For a great and readable history of both Bowlby and Spitz (and from which I draw significant history in this section), see Lydia Denworth, *Friendship: The Evolution, Biology, and Extraordinary Power of Life's Fundamental Bond* (New York: W.W. Norton & Company, 2020).

13 See John B. Watson, *Psychological Care of Infant and Child* (New York: W.W. Norton & Company, 1928), 69, 81–82, 87.

14 René A. Spitz, "Hospitalism: An Inquiry into the Genesis of Psychiatric Conditions in Early Childhood," *The Psychoanalytic Study of the Child* 1, No. 1 (1945), 54–55.

15 Ibid., 68.

16 René A. Spitz, "Hospitalism: A Follow-Up Report on Investigation Described in Volume I, 1945," *The Psychoanalytic Study of the Child* 2, No. 1 (1946), 115.

17 Bowlby, "The Nature of the Child's Tie," 362.

18 See again a great synopsis by Denworth, *Friendship*.

19 As quoted by David Brooks, *The Second Mountain: The Quest for a Moral Life* (New York: Random House, 2019), xxvi.

20 See Susan Brink, *The Fourth Trimester: Understanding, Protecting, and Nurturing an Infant Through the First Three Months* (Berkeley: University of California Press, 2013).

21 David F. Bjorkland, "The Role of Immaturity in Human Development," *Psychological Bulletin* 122, No. 2 (1997), 153–169. See also Zachary Cofran, "Brain Size Growth in Wild and Captive Chimpanzees (Pan troglodytes)," *American Journal of Primatology* 80 (2018), e22876.

22 Brink, *The Fourth Trimester.* See also, Denworth, *Friendship*.

23 Lydia Denworth provides a great description of this phenomenon. As she writes, even though infants are not very interactive in the first months of life, "an enormous amount of social infrastructure is being laid from the start. The

senses—vision, hearing, touch, smell, and taste—are conduits through which a baby takes in the details of her new environments and conveys them to her brain, which is preprogrammed with a preference for social interaction. . . . Faces, voices, and loving caresses." See Denworth, *Friendship*, 45. In his own words, Bowlby describes this phenomenon as follows: "When he is born, an infant is far from being a [blank slate]. On the contrary, . . . he [is] equipped with a number of behavioural systems ready to be activated. . . . Among these systems . . . are . . . some that provide the building-bricks for the later development of attachment. Such, for example, are the primitive systems [of] crying, sucking, clinging, and orientation. To these are added, only a few weeks later, smiling and babbling. . . . From the start there is a marked bias to respond in special ways to several kinds of stimuli that commonly emanate from a human being—the auditory stimuli arising from a human voice, the visual stimuli arising from a human face, and the tactile . . . stimuli arising from human arms and body. From these small beginnings spring all the highly discriminating and sophisticated systems that in later infancy and childhood—indeed for the rest of life—mediate" relationships with other people. See John Bowlby, *Attachment and Loss: Volume I: Attachment* (New York: Basic Books, 1969), 265–266.

24 Bowlby, *Attachment and Loss*, 207–208.

25 Martin E. P. Seligman, *Flourish: A Visionary New Understanding of Happiness and Well-Being* (New York: Atria, 2011), 17.

26 Ibid.

27 Ibid., 20 (italics in original). Two other psychologists, Harry Reis and Shelly Gable, come to essentially the same conclusion: "The weight of evidence is, in short, so compelling that one commentator referred to the association between relationships and well-being as a 'deep truth.' Virtually all reviews of this vast and diverse literature have concluded similarly" that good "relationships are . . . perhaps the most important . . . source of life satisfaction and well-being." See Harry T. Reis and Shelly L. Gable, "Toward a Positive Psychology of Relationships," in *Flourishing: Positive Psychology and the Life Well-Lived*, eds. Corey L. M. Keyes and Jonathan Haidt (Washington: American Psychological Association: 2003), 129, 130.

28 Notably, humans are incredibly adaptable to most adverse circumstances that might affect well-being. We can achieve relative satisfaction and well-being in a host of difficult situations. This is a phenomenon known as hedonic adaptation. In one famous study, researchers examined the happiness and well-being of two groups of people: lottery winners and accident victims who were left paralyzed. In terms of how much pleasure each experienced on

a day-to-day basis (from ordinary events like hearing a funny joke, talking with a friend, etc.), there was no difference between the groups. It's true that the accident victims were slightly less happy overall compared to the lottery winners; however, they had learned to adapt to very difficult circumstances and find pleasure in the simple things of life. The initial thrill of winning the lottery also wore off. (See Philip Brickman, Dan Coates, and Ronnie Janoff-Bulman, "Lottery Winner and Accident Victims: Is Happiness Relative?" *Journal of Personality and Social Psychology* 36, No. 8 (1978), 917–927.) Economist Robert Frank gives another example of the human ability to adapt: "As a young man fresh out of college, I served as a Peace Corps Volunteer in rural Nepal. My one-room house had no electricity, no heat, no indoor toilet, no running water. The local diet offered little variety and virtually no meat. . . . Yet, although my living conditions in Nepal were a bit startling at first, the most salient feature of my experience was how quickly they came to seem normal. Within a matter of weeks, I lost all sense of impoverishment." (As quoted in David G. Myers, "The Funds, Friends, and Faith of Happy People," *American Psychologist* 55, No. 1 (2000), 61.) The study on accident victims and lottery winners as well as Frank's experience both attest that humans are incredibly capable of adapting to and even thriving amid adversity. But relationships are the exception to this. Humans do not adapt to unhealthy and harmful relationships.

29 Angus Campbell, Philip Converse, and Willard Rodgers, *The Quality of American Life: Perceptions, Evaluations, and Satisfactions* (New York: Russell Sage Foundation, 1976).

30 Norval D. Glenn and Charles N. Weaver, "The Contribution of Marital Happiness to Global Happiness," *Journal of Marriage and the Family* 43, No. 1 (1981), 161.

31 Dorie Giles Williams, "Gender, Marriage, and Psychological Well-Being," *Journal of Family Issues* 9, No. 4 (1988), 452–468.

32 Walter Gove, Michael Hughes, and Carolyn Style, "Does Marriage Have Positive Effects on the Psychological Well-Being of the Individual?" *Journal of Health and Social Behavior* 24, No. 2 (1983), 122–131.

33 David G. Blanchflower and Andrew J. Oswald, "International Happiness: A New View on the Measure of Performance," *Academy of Management Perspectives* 25, No. 1 (Feb. 2011), 6–22. The next strongest predictor in this data tabulation was unemployment, which affected happiness in the opposite direction as marriage. See Table 1.

34 Steven Stack and J. Ross Eshleman, "Marital Status and Happiness: A 17-Nation Study," *Journal of Marriage and Family* 60, No. 2 (May 1998), 527–536.

35 Arne Mastekaasa, "Marital Status, Distress, and Well-Being: An International Comparison," *Journal of Comparative Family Studies* 25, No. 2 (Summer 1994), 183–205.

36 Kristen Schultz Lee and Hiroshi Ono, "Marriage, Cohabitation, and Happiness: A Cross-National Analysis of 27 Countries," *Journal of Marriage and Family* 74 (Oct. 2012), 953–972.

37 In a recent analysis of data spanning eighteen years and representing approximately 30,000 individuals, Shawn Grover and John F. Helliwell track individuals before and after marriage. Among those who eventually marry, Grover and Helliwell statistically control for premarital levels of well-being in an effort to account for the selection of happier people into marriage. What they find is that people with higher levels of well-being are more likely to get married. But this only accounts for a relatively small proportion of the effect that marriage has on well-being. In other words, the analysis by Grover and Helliwell suggest that most of the happiness difference between married people and single people is due to marriage causing people to be happier. See Shawn Grover and John F. Helliwell, "How's Life at Home? New Evidence on Marriage and the Set Point for Happiness," *Journal of Happiness Studies* 20 (2019), 373–390. For a great starting point to the large body of literature suggesting that marriage has causal effects on health and happiness for the better, see Tyler J. VanderWeele, "What the *New York Times* Gets Wrong About Marriage, Health, and Well-Being," *Institute for Family Studies*, May 30, 2017: https://ifstudies.org/blog/what-the-new-york-times-gets-wrong-about-marriage-health-and-well-being.

38 Allan V. Horwitz, Helene Raskin White, and Sandra Howell-White, "Becoming Married and Mental Health: A Longitudinal Study of a Cohort of Young Adults," *Journal of Marriage and the Family* 58 (Nov. 1996), 895.

39 Nadine F. Marks and James D. Lambert, "Marital Status Continuity and Change Among Young and Midlife Adults: Longitudinal Effects on Psychological Well-being," *Journal of Family Issues* 19, No. 6 (Nov. 1998), 652–686.

40 Linda J. Waite and Maggie Gallagher, *The Case for Marriage: Why Married People Are Happier, Healthier, and Better Off Financially* (New York: Broadway Books, 2000), 68.

41 First quote from Michael Argyle, "Causes and Correlates of Happiness," in *Well-Being: The Foundations of Hedonic Psychology*, eds. Daniel Kahneman, Edward Diener, and Norbert Schwarz (New York: Russell Sage Foundation, 1999), 361. The second quote is from Waite and Gallagher, *The Case for Marriage*, 77.

42 Waite and Gallagher, *The Case for Marriage.*

43 David G. Myers, "The Funds, Friends, and Faith of Happy People," *American Psychologist* 55, No. 1 (2000), 56–67.

44 A wise leader once declared: "In terms of your happiness, in terms of the matters that make you proud or sad, nothing . . . will have so profound an effect on you as the way your children turn out. You will either rejoice and boast of their accomplishments or you will weep, head in hands, bereft and forlorn, if they become a disappointment or an embarrassment to you." See Gordon B. Hinckley, "Great Shall Be the Peace of Thy Children," *General Conference Address of The Church of Jesus Christ of Latter-day Saints*, Oct. 7, 2000.

45 Laura Silver, Patrick Van Kessel, Christine Huang, Laura Clancy, and Sneha Gubbala, "What Makes Life Meaningful? Views from 17 Advanced Economies," *Pew Research Center*, Nov. 18, 2021, available at https://www.pewresearch.org/global/2021/11/18/what-makes-life-meaningful-views-from-17-advanced-economies/.

46 "Where Americans Find Meaning in Life," *Pew Research Center*, Nov. 20, 2018, available at https://www.pewresearch.org/religion/2018/11/20/where-americans-find-meaning-in-life/.

47 Daniel Gilbert, *Stumbling on Happiness* (New York: Alfred A. Knopf, 2006), 222.

48 Wendy Wang, "Parents' Time with Kids More Rewarding Than Paid Work—And More Exhausting," *Pew Research Center*, Oct. 8, 2013, available at https://www.pewresearch.org/social-trends/2013/10/08/parents-time-with-kids-more-rewarding-than-paid-work-and-more-exhausting/.

49 Roy Baumeister, Kathleen D. Vohs, Jennifer L. Aaker, and Emily N. Garbinsky, "Some Key Differences Between a Happy Life and a Meaningful Life," *Journal of Positive Psychology* 8, No. 6 (2013), 505–516.

50 David G. Blanchflower and Andrew E. Clark, "Children, Unhappiness and Family Finances," *Journal of Population Economics* 34 (2021), 625–653.

51 As reported in Joe Pinsker, "It Isn't the Kids. It's the Cost of Raising Them," *The Atlantic*, Feb. 27, 2019, available at https://www.theatlantic.com/family/archive/2019/02/cost-raising-kids-parents-happiness/583699/.

52 Jennifer Glass, Matthew A. Andersson, and Robin W. Simon, "Parenthood and Happiness: Effects of Work-Family Reconciliation Policies in 22 OECD Countries," *American Journal of Sociology* 122, No. 3 (Nov. 2016), 886–929.

53 See Bryan Caplan, *Selfish Reasons to Have More Kids* (New York: Basic Books, 2011), 14–15.

54 See, for example, Jennifer Senior, *All Joy and No Fun: The Paradox of Modern Parenthood* (New York: HarperCollins, 2014).

55 Adam Smith, *An Inquiry Into the Nature and Causes of the Wealth of Nations* (London, 1776), 181.

56 Armand M. Nicholi, "The Impact of Family Dissolution on the Emotional Health of Children and Adolescents," in *When Families Fail . . . the Social Costs*, ed. Bryce J. Christensen (Lanham, Md.: University Press of America, 1991), 38.

57 Carl Rogers, as quoted in Michael Wallach and Lise Wallach, "How Psychology Sanctions the Cult of the Self," *Washington Monthly*, Feb. 1985, 47.

58 This is in line with the so-called self-esteem fad that gained momentum in the 1970s and has culminated in our current culture of self-obsession.

59 See, for instance, Robert Putnam, *Bowling Alone: The Collapse and Revival of American Community* (New York: Simon & Schuster, 2000).

60 See Brooks, *The Second Mountain*.

61 For example, the psychologist David Myers summarized our achievements in well-being from the latter half of the 20th century: "Compared with their grandparents, today's young adults have grown up with much more affluence, slightly less happiness, and much greater risk of depression and assorted social pathologies. . . . The more people strive for extrinsic goals . . . the more numerous their problems and the less robust their well-being." See Myers, "The Funds, Friends, and Faith of Happy People," 61. The psychologists Michael and Lise Wallach from Duke University similarly concluded that the cultural shift of the last half of the 20th century "has legitimized a self-centered view of humanity that not only damages society but also leads people away from the very happiness they seek." See Wallach and Wallach, "How Psychology Sanctions the Cult of the Self," 49.

62 Luke 17:33 (King James Version).

63 See Lottie Bullens, Jens Forster, Frenk van Harreveld, and Nira Liberman, "Self-Produced Decisional Conflict Due to Incorrect Metacognitions," in *Cognitive Consistency: A Fundamental Principle in Social Cognition*, eds. Bertram Gawronski and Fritz Strack (New York: Guilford Press, 2012), 285–304. See also Barry Schwartz, *The Paradox of Choice* (New York: HarperCollins, 2004).

64 Daniel T. Gilbert and Jane E. J. Ebert, "Decisions and Revisions: The Affective Forecasting of Changeable Outcomes," *Journal of Personality and Social Psychology* 82, No. 4 (2002), 511.

65 See, for instance, Kristen Schultz Lee and Hiroshi Ono, "Marriage, Cohabitation, and Happiness: A Cross-National Analysis of 27 Countries," *Journal of Marriage and Family* 74 (Oct. 2012), 953–972.

66 David Brooks, "The Age of Possibility," *New York Times*, Nov. 16, 2012, A35, nytimes.com/2012/11/16/opinion/brooks-the-age-of-possibility.html.

67 Clayton M. Christensen, James Allworth, and Karen Dillon, *How Will You Measure Your Life?* (New York: HarperCollins, 2012), 80.

Chapter 10: The Good Society

1 Ernest R. Sandeen, "John Humphrey Noyes as the New Adam," *Church History* 40, No. 1 (1971), 82.
2 Ibid.
3 Ibid.
4 Noyes once said, after being dismissed from Yale and expelled from the preaching community, "I have taken away their license to sin, and they keep on sinning. So, though they have taken away my license to preach, I shall keep on preaching." John H. Noyes, *Confessions of John H. Noyes. Part I. Confession of Religious Experience: Including a History of Modern Perfectionism* (Oneida, N.Y.: Leonard & Company Printers, 1849), 27.
5 Lawrence Foster, "Free Love and Feminism: John Humphrey Noyes and the Oneida Community," *Journal of the Early Republic* 1, No. 2 (Summer 1981), 173.
6 Sandeen, "John Humphrey Noyes," 83.
7 Spencer Klaw, *Without Sin: The Life and Death of the Oneida Community* (New York: Allen Lane, Penguin, 1993), 170.
8 Ibid., 170–171.
9 Ibid., 144.
10 Ibid., 143.
11 Ibid.
12 Sigmund Freud, *Civilization and Its Discontents,* transl. David McLintock (New York: Penguin, 2004), 43, 77.
13 David Brooks, *The Second Mountain: The Quest for a Moral Life* (New York: Random House, 2019), 308.
14 Christakis conducts an in-depth analysis of several of these Utopian experiments, including the Shakers and the kibbutzim in Israel. I draw heavily on his conclusions about how subverting the most biologically primed sources of love and affection can lead to an unstable society. See *Blueprint: The Evolutionary Origins of a Good Society.*
15 Referring to the two extreme approaches to sexuality (complete celibacy versus a very liberal and promiscuous ethos of sexuality) seen in some of these communal experiments, Nicholas Christakis notes: "Both of these strategies shared the common aim of subverting the institution of marriage and eroding any deep personal connections between pairs of individuals; the objective of these strategies was to foster a sense of connection to the group as a whole.

This is the reason that many communities also attempted to break down the nuclear family through communal child care and separate living quarters for parents and children. . . . Such efforts almost always fail because they subvert our species' pre-wired instinct for love." *Blueprint*, 98.

16 U.S. Department of Justice, *Crimes in the Unites States, 2011*, Table 42, accessed May 5, 2022, https://ucr.fbi.gov/crime-in-the-u.s/2011/crime-in-the-u.s.-2011/tables/table-42.

17 U.S. Department of Justice, *Correctional Population in the United States, 2010*, December 2011, https://bjs.ojp.gov/content/pub/pdf/cpus10.pdf.

18 David Gutmann, "Parenthood: A Key to Comparative Study of the Life Cycle," in *Life-Span Developmental Psychology*, eds. Nancy Datan and Leon H. Ginsberg (Cambridge, Mass.: Academic Press, 1975), 167–184.

19 While this statement is true of divorce since the mid-1900s, there has been a recent increase in true shared custody following divorce, perhaps as we have learned that fathers are critical to the development of children. See Daniel R. Meyer, Marcia J. Carlson, and Md Moshi Ul Alam, "Increases in Shared Custody after Divorce in the United States," *Demographic Research* 46, No. 38, (2022), 1137–1162.

20 As quoted in Alyse ElHage, "How Marriage Makes Men Better Fathers," *Institute for Family Studies*, June 19, 2015, https://ifstudies.org/blog/how-marriage-makes-men-better-fathers.

21 Solomon Jones, as quoted in Alyse ElHage, "How Marriage Makes Men Better Fathers."

22 Nicholas Zill, Donna Ruane Morrison, and Mary Jo Coiro, "Long-Term Effects of Parental Divorce on Parent-Child Relationships, Adjustment, and Achievement in Young Adulthood," *Journal of Family Psychology* 7, No. 1 (1993), 91–103.

23 Jay Belsky, Lise Youngblade, Michael Rovine, and Brenda Volling, "Patterns of Marital Change and Parent-Child Interaction," *Journal of Marriage and Family* 53, No. 2 (1991), 487–498.

24 Frank F. Furstenberg, "Good Dads—Bad Dads: Two Faces of Fatherhood," in *The Changing American Family and Public Policy*, ed. Andrew Cherlin (Washington: Urban Press Institute, 1988), 201. Quote from 201, 193–218 overall.

25 W. Bradford Wilcox, *Why Marriage Matters: Thirty Conclusions from the Social Sciences*, 3rd Edition (New York: Institute for American Values, 2011).

26 See Frank F. Furstenberg Jr. and Andrew Cherlin, *Divided Families: What Happens to Children When Parents Part* (Cambridge, Mass.: Harvard University Press, 1991).

27 See Carolyn Pape Cowan and Philip A. Cowan, "Men's Involvement in Parenthood: Identifying the Antecedents and Understanding the Barriers," in *Men's Transitions to Parenthood: Longitudinal Studies of Early Family Experience*, eds. Phyllis W. Berman and Frank A. Pedersen (New York: Lawrence Erlbaum Associates, 1987), 145–174.

28 Popenoe, *Life Without Father*, 75.

29 Waite and Gallagher, *The Case for Marriage.*

30 Robert J. Sampson, John H. Laub, and Christopher Wimer, "Does Marriage Reduce Crime? A Counterfactual Approach to Within-Individual Causal Effects," *Criminology* 44, No. 3 (2006), 465–508.

31 Waite and Gallagher, *The Case for Marriage.*

32 Popenoe, *Life Without Father*, 77.

33 See Kristen Harknett and Arielle Kuperberg, "Education, Labor Markets, and the Retreat from Marriage," *Social Forces* 90, No. 1 (Sept. 1, 2011), 41–63. The relationship between marriage and the economy is complex and debate remains as to how much economic factors (and for which groups) influence marriage and family structure. See Daniel T. Litcher, Diane K. McLaughlin, and David C. Ribar, "Economic Restructuring and the Retreat from Marriage," *Social Science Research* 31, No. 2 (June 2002), 230–256.

34 W. Bradford Wilcox, "Don't Be a Bachelor: Why Married Men Work Harder, Smarter and Make More Money," *Washington Post*, April 2, 2015, https://www.washingtonpost.com/news/inspired-life/wp/2015/04/02/dont-be-a-bachelor-why-married-men-work-harder-and-smarter-and-make-more-money/.

35 When a bachelor friend complained that he was too poor to marry, Benjamin Franklin responded: "A single man has not nearly the value he would have in that state of union [marriage]. He is an incomplete animal. He resembles the odd half of a pair of scissors. If you get a prudent healthy wife, your industry in your profession, with her good economy, will be a fortune sufficient." Benjamin Franklin, "Old Mistresses Apologue," *Franklin Papers*, June 25, 1745, available at https://founders.archives.gov/documents/Franklin/01-03-02-0011.

36 See Waite and Gallagher, *The Case for Marriage*, Chapter 4.

37 Sampson, Laub, and Wimer, "Does Marriage Reduce Crime? A Counterfactual Approach to Within-Individual Causal Effects," 465–508. See also Waite and Gallagher, *The Case for Marriage.*

38 Donna K. Ginther and Madeline Zavodny, "Is the Male Marriage Premiums Due to Selection? The Effect of Shotgun Weddings on the Return to Marriage," *Journal of Population Economics* 14, No. 2 (June 2001), 313–328; see also Sanders Korenman and David Neumark, "Does Marriage Really Make

Men More Productive?" *Journal of Human Resources* 26, No. 2 (Spring 1991), 282–307.

39 Waite and Gallagher, *The Case for Marriage.*

40 For a review, see Chris M. Wilson and Andrew J. Oswald, "How Does Marriage Affect Physical and Psychological Health? A Survey of the Longitudinal Evidence," Discussion Paper No. 1610 (May 2005), available at https://papers.ssrn.com/sol3/papers.cfm?abstract_id=735205. See also Waite and Gallagher, *The Case for Marriage.*

41 Kate Antonovics and Robert Town, "Are All the Good Men Married? Uncovering the Sources of the Marital Wage Premium," *The American Economic Review* 94, No. 2 (May 2004), 317–321. Married men are also substantially more likely to be happy compared to their identical twins who are single.

42 See, for example, Matthijs Kalmijn, "The Effects of Divorce on Men's Employment and Social Security Histories," *European Journal of Population* 21 (2005), 347–366.

43 Popenoe, *Life Without Father.*

44 Joseph P. Shapiro and Joannie M. Schrof, "Honor Thy Children," *US News and World Report*, 118, No. 8 (Feb. 27, 1995), 38–46.

45 One criticism of this argument is that it would have been very costly for these women to "civilize" these men were they to have married them. Many of these men may have had serious drug, alcohol, or other problems. There may be some truth to this, but, as discussed in this chapter, the very act of demonstrating to the community his commitment to a woman (through marriage) can profoundly influence a man's behavior for the better. The data suggest it certainly would be better for the women to have married the men than to have cohabitated with them.

46 John Snarey, *How Fathers Care for the Next Generation: A Four-Decade Study* (Cambridge, Mass.: Harvard University Press, 1993), 118.

47 See David Popenoe, *War Over the Family* (New Brunswick, N.J.: Transaction Publishers, 2005), 119.

48 Blankenhorn, *Fatherless America*, 25.

49 Kathleen O'Brien, "David Popenoe: The Marriage Dude." *Inside New Jersey*, Feb. 11, 2007, available at https://www.nj.com/iamnj/2007/02/david_popenoe.html.

50 *Families First: Report of the National Commission on America's Urban Families* (Washington: U.S. Government Printing Office, 1993), iii.

51 Popenoe, *War Over the Family*, 120.

52 O'Brien, "David Popenoe."

53 Blankenhorn, *Fatherless America*, 22.

54 See Popenoe, *Life Without Father*; Blankenhorn, *Fatherless America*.

55 I should note that there are some confounding effects likely going on here. In the last fifty years, marriage has increasingly become an institution of the wealthy and educated. People with higher socioeconomic status are much more likely to marry (and remain married) compared to those of less means and education. There has been a marked retreat from marriage not only among the poor and working class, but also among the middle class. Nonetheless, there do seem to be ways that family structure influences stability beyond confounding effects. Marriage serves as a sort of commitment device, tying men and women together and to their children, that has positive effects on family stability beyond these confounding variables.

56 Francisco Perales, Sarah E. Johnson, Janeen Baxter, David Lawrence, and Stephen R. Zubrick, "Family Structure and Childhood Mental Disorders: New Findings from Australia," *Social Psychiatry and Psychiatric Epidemiology* 52 (2017), 423–433.

57 Dana Lizardi, Ronald G. Thompson, Katherine Keyes, and Deborah Hasin, "Parental Divorce, Parental Depression, and Gender Differences in Adult Offspring Suicide Attempt," *Journal of Nervous and Mental Disease* 197, No. 12 (Dec. 2009), 899–904.

58 Sara McLanahan and Gary Sandefur, *Growing Up with a Single Parent: What Hurts, What Helps* (Cambridge, Mass.: Harvard University Press, 1994), 1–2.

59 Deborah A. Dawson, "Family Structure and Children's Health and Well-Being: Data from the 1988 National Health Interview Survey on Child Health," *Journal of Marriage and the Family* 53, No. 3 (Aug. 1991), 573–584.

60 McLanahan and Sandefur, *Growing Up with a Single Parent*, 1–2.

61 See Trudi J. Renwick, *Poverty and Single Parent Families* (New York: Garland Publishing, 1998). For a more up to date reference, see also Robert I. Lerman and W. Bradford Wilcox, *For Richer, For Poorer: How Family Structures Economic Success in America* (Washington: American Enterprise Institute, 2014).

62 David T. Ellwood, *Poor Support: Poverty in the American Family* (New York: Basic Books, 1988), 46. Although this quote is several decades old, its basic premise seems to have held up over time. A report from the National Center for Education Statistics from 2020 showed that children living in two-parent households are much less likely to be in poverty compared to children living in single-parent households. See National Center for Education Statistics, "Characteristics of Children's Families," May 2022 (last updated), https://nces.ed.gov/programs/coe/indicator/cce.

63 Pearl L. H. Mok, Aske Astrup, Matthew J. Carr, Sussie Antonsen, Roger T. Webb, and Carsten B. Pedersen, "Experience of Child-Parent Separation and Later Risk of Violent Criminality," *American Journal of Preventive Medicine* 55, No. 2 (Aug. 2018), 178–186.

64 Henry B. Biller, *Fathers and Families: Paternal Factors in Child Development* (Westport, Conn.: Auburn House, 1993), 2.

65 Bruce J. Ellis, John E. Bates, Kenneth A. Dodge, David M. Fergusson, L. John Horwood, Gregory S. Pettit, and Lianne Woodward, "Does Father Absence Place Daughters at Special Risk for Early Sexual Activity and Teenage Pregnancy?" *Child Development* 74, No. 3 (May–June 2003), 801–821.

66 Raj Chetty, John N. Friedman, Nathaniel Hendren, Maggie R. Jones, and Sonya R. Porter, "The Opportunity Atlas: Mapping the Childhood Roots of Social Mobility," NBER Working Paper No. 25147, Oct. 2018, revised Feb. 2020.

67 W. Bradford Wilcox, Joseph Price, and Robert I. Lerman, "Strong Families, Prosperous States: Do Healthy Families Affect the Wealth of States?" The Home Economics Project, a project of the American Enterprise Institute and the Institute for Family Studies, 2015.

68 Chetty and colleagues, "The Opportunity Atlas," 2018.

69 Adoption represents an extremely unselfish act on the part of the birth mother as well as the adoptive parents. How does adoption influence childhood outcomes? It's a little complicated, and much depends upon the circumstances surrounding the adoption. Children who are adopted earlier in life experience better outcomes than those who spend significant time in foster care or in the care of the government. When adoption occurs, the child has usually already experienced traumatic events, such as abuse, neglect, and the dissolution of their attachment with the mother or another caregiver. Adopted children are also more likely to have experienced adverse circumstances while still in the womb (including exposure to drugs or alcohol and poorer nutrition). These early adversities are difficult to overcome. Consequently, adopted children usually have more behavioral and emotional problems compared to their peers who were not adopted. Research has shown, however, that following adoption, the behavioral and emotional problems of children often improve significantly. This is especially true if the adoptive parents foster a relationship of warmth and empathy. Notwithstanding the challenges experienced by adopted children, most scholars agree that adoption represents a better option for the child than remaining with the birth mother or staying in the care of the state. See Amy L. Paine, Oliver Perra, Rebecca Anthony, and Katherine H. Shelton, "Charting the Trajectories of Adopted Children's Emotional and Behavioral Problems: The Impact

of Early Adversity and Postadoptive Parental Warmth," *Development and Psychopathology* 33 (2021), 922–936. See also Margaret F. Keil, Adela Leahu, Megan Rescigno, Jennifer Myles, and Constantine A. Stratakis, "Family Environment and Development in Children Adopted from Institutionalized Care," *Pediatric Research* 91 (2022), 1562–1570.

70 Some scholars might object to my use of the example of child abuse, as there is some evidence that rates of child abuse have decreased in the United States in recent decades. However, the quality of data collected in the 1950s and 1960s (before divorce started becoming much more prevalent) is lower than the quality of data in recent decades, making comparisons and trends analyses spanning this time period problematic. I will point out that there is evidence that the rate of adverse childhood experiences, which would include abuse as well as other hardships (such as neglect), has increased significantly in the last seventy years. See Philip M. Hughes, Tabitha L. Ostrout, Mónica Pèrez, and Kathleen C. Thomas, "Adverse Childhood Experiences Across Birth Generation and LGBTQ+ Identity, Behavioral Risk Factor Surveillance System, 2019," *American Journal of Public Health* 112, No. 4 (2022), 662–670.

71 Popenoe, *Life Without Father.*

72 Lawrence M. Berger, Christina Paxson, and Jane Waldfogel, "Mothers, Men, and Child Protective Services Involvement," *Child Maltreatment* 14, No. 3 (Aug. 2009), 263–276; see also Aruna Radhakrishna, Ingrid E. Bou-Saada, Wanda M. Hunter, Diane J. Catellier, and Jonathan B. Kotch, "Are Father Surrogates a Risk Factor for Child Maltreatment?" *Child Maltreatment* 6, No. 4 (Nov. 2001), 281–289.

73 Leslie Margolin, "Child Abuse by Mothers' Boyfriends: Why the Overrepresentation?" *Child Abuse & Neglect* 16 (1992), 541–551. See also William J. Oliver, Lawrence R. Kuhns, and Elaine S. Pomeranz, "Family Structure and Child Abuse," *Clinical Pediatrics* 45, No. 2 (March 2006), 111–118.

74 Rebecca L. Burch and Gordon G. Gallup Jr., "Abusive Men Are Driven by Paternal Uncertainty," *Evolutionary Behavioral Sciences* 14, No. 2 (April 2020), 197–209.

75 Popenoe, *Life Without Father*, 65–73.

76 The risk of domestic violence, for instance, is lower among married women than among divorced or cohabitating women. See Waite and Gallagher, *The Case for Marriage.*

Chapter 11: Final Thoughts

1 Wilson, *On Human Nature*, 1.

2 W. Bradford Wilcox and Kathleen Kline, *Gender and Parenthood: Biological and Social Scientific Perspectives* (New York: Columbia University Press, 2013), 3.

3 See "Few Americans Blame God or Say Faith Has Been Shaken Amid Pandemic, Other Tragedies," *Pew Research Center*, Nov. 23, 2021, available at https://www.pewresearch.org/religion/2021/11/23/few-americans-blame-god-or-say-faith-has-been-shaken-amid-pandemic-other-tragedies/.

4 Darwin, *The Origin of Species*, 384.

5 Ibid., 417 (italics added).

6 For the record, Darwin later wrote to a friend that he regretted having "truckled to public opinion" and used such language in later editions. See Charles Darwin to Joseph Hooker, March 29, 1863, in the *Darwin Correspondence Project*, https://www.darwinproject.ac.uk/letter/DCP-LETT-4065.xml.

Index

Note: page numbers in *italics* indicate figures.